*Classic Geology in Europe 12*

# Almeria

# CLASSIC GEOLOGY IN EUROPE

For details of these and other earth sciences titles from Dunedin
Academic Press see www.dunedinacademicpress.co.uk

*Classic Geology in Europe 12*

# Almeria

**Adrian M. Harvey**
Emeritus Professor of Geomorphology, School of Environmental
Sciences, University of Liverpool

**Anne E. Mather**
School of Geography, Earth and Environmental Sciences,
Plymouth University

EDINBURGH ◆ LONDON

Published in the United Kingdom by
Dunedin Academic Press Ltd
*Head Office:*
Hudson House, 8 Albany Street, Edinburgh, EH1 3QB
*London Office:*
352 Cromwell Tower, Barbican, London, EC2Y 8NB

**bitlit**

A **free** eBook edition is available
with the purchase of this print book.

CLEARLY PRINT YOUR NAME ABOVE IN UPPER CASE
**Instructions to claim your free eBook edition:**
1. Download the BitLit app for Android or iOS
2. Write your name in **UPPER CASE** on the line
3. Use the BitLit app to submit a photo
4. Download your eBook to any device

ISBNs:
978-1-78046-037-6 (paperback)
978-1-78046-527-2 (epub)
978-1-78046-528-9 (Kindle)

© 2015 Adrian M. Harvey & Anne E. Mather

The right of Adrian Harvey and Anne Mather to
be identified as the authors of this book has been
asserted by them in accordance with sections 77 &
78 of the Copyright, Designs and Patents Act
1988

British Library Cataloguing in Publication Data
A catalogue record for this book is available from the British Library

Typeset by Makar Publishing Production, Edinburgh
Printed in Poland by Hussar Books

# Contents

# About the authors

Adrian Harvey graduated in 1962 with a BSc degree in Geography (with Geology), then in 1967 with a PhD in Geomorphology, both degrees from University College, London. From 1965 he was on the staff in the Department of Geography (latterly the School of Environmental Sciences) at the University of Liverpool until retirement as Professor of Geomorphology in 2005. His research has focussed on fluvial system dynamics, especially on active systems but also on arid regions. Since retirement he has been Emeritus Professor of Geomorphology in the University of Liverpool.

Anne Mather graduated in 1984 with a BSc degree in Physical Geography and Geology at the University of Hull and after a three-year stint in the Geoscience industry returned to academia to complete a PhD in fluvial system sedimentology and sedimentary basin dynamics (Almeria) at the University of Liverpool in 1991. That research evolved to encompass long-term controls on the geomorphology of tectonically active arid regions. After her PhD she worked for a short while at the University of Worcester before moving to Plymouth, where she is now an Associate Professor (Reader) in the School of Geography, Earth and Environmental Sciences.

.

# Preface

This book deals with the geology and geomorphology of Almeria province in southeast Spain. At the southeast 'corner' of the Iberian peninsula, Almeria is a classic area for south European/Mediterranean Neogene and Quaternary geology, including geomorphology. The book is not aimed at the specialist research scientist, but at a more general readership, including students and members of the general public with an interest in the earth sciences. We assume some knowledge of geology, but present several appendices that deal with general principles as well as a glossary, defining particular terms.

Since the Mesozoic (*see* Appendix 1 for nomenclature related to the geological timescale) SE Spain has been affected by the interaction between the European and African tectonic plates. Initially the collision emplaced a series of nappes, thrust from the south, creating the major structures of the Betic mountain chain. The main rock groups involved in these structures are mostly a wide range of metamorphic rocks that today form the basement to the region. During the late-Tertiary (the Neogene, see Appendix 2), the overall compressional tectonic regime was replaced by lateral shear (the trans-Alboran shear zone). This created the 'basin-and-range' terrain of the region today, defined in part by a series of major strike-slip fault systems. The Neogene rocks include important volcanic rocks. The main body of these was extruded in what is now the Alboran Sea, but has been brought into the area along one of the strike-slip fault systems. The sedimentary basins between the uplifted mountain blocks are largely defined by the strike-slip faults. During the Neogene, the basins themselves underwent first marine, then terrestrial sedimentation. They preserve within them an extraordinary range of sedimentary rocks including evaporitic gypsum related to the 'Messinian salinity crisis', a period when the Mediterranean Sea desiccated. Differential epeirogenic uplift of the mountains and amongst the various basins over the same timescale created the outlines of the modern topography. The region is still tectonically active, with ongoing mountain uplift and deformation along the basin-bounding faults and within the sedimentary basins.

The geomorphology of the region reflects not only the regional tectonics, but also the changing climates of the Quaternary (*see* Appendix 3), and particularly the Pleistocene and modern semi-arid climates. Since emergence from the late Neogene seas, the region has developed in a terrestrial environment. Differential epeirogenic uplift determined the main outlines of the drainage. The river systems have incised into the underlying basin-fill rocks – incidentally, creating superb exposures of the Neogene sedimentary sequences. Drainage patterns have been modified by river capture, affecting local base levels and incision rates. Areas of rapid incision, especially into the more resistant rocks, are marked by spectacular canyons. The regional incisional sequence has been punctuated by periods of climatically-driven sediment excess, producing a suite of river terraces within an overall incisional landscape. At the same time, in zones away from the incising drainage network large alluvial fans were deposited. This is one of the best areas in Europe to study Quaternary and modern alluvial fans. Indeed, alluvial fans, though not exclusive to drylands, are one of the characteristic landforms of dryland mountain regions. This region is the driest in Europe and its climate influences the modern geomorphic processes. The vegetation cover is relatively thin, allowing erosion processes to be effective. On the softer rocks, deeply dissected badlands are common, especially where the long human occupation of the region has reduced vegetation cover and accelerated erosion. The modern river channels, too, are characteristic of dryland mountain environments, often exhibiting braided patterns. Another aspect of the dryland environment is the exceptional development of pedogenic calcretes, forming resistant caprocks over weaker underlying sediments. One unusual aspect of the modern geomorphology relates to the legacy of the Messinian salinity crisis. There are exceptional gypsum karst landscapes with a wide range of 'gypkarst' surface features and cave systems that allow the study of gypsum karst features below as well as above ground. Finally, the coast needs to be mentioned. Although there is some tourist development, much of the coast is natural, and includes a wide range of beaches and cliffed coasts. Preserved at a number of sites along the coast are raised beaches dating from past interglacial sea-level highstands, providing a framework for dating the Pleistocene sequence of landform evolution.

In summary, the region exhibits superb structural geology (especially the fault systems), a complete Neogene sedimentary sequence (itself rare) involving an enormous range of sedimentary environments, and classic dryland geomorphology. Exposure of the sedimentary sequences is excellent. The area is spectacular and the landform assemblage includes a wide range of erosional and depositional landscapes. Furthermore, the region enables linkages to be made between the several disciplines of geodynamics. One cannot interpret the Neogene sequence without considering the evolving tectonics and the contemporaneous geomorphology. One cannot fully interpret the modern geomorphology without considering the landscape as a development from the Neogene tectonic and sedimentary sequences. For these reasons, and because of our own expertise, the emphases in the book are primarily on the Neogene and Quaternary, on the sedimentary sequences and the geomorphology. We do, however, put these into the wider and longer-term geological context.

The area is frequently used for university field classes in geology and geography from the UK and elsewhere, as well as having potential for 'geo-tourism'. We have written this book with these two types of readership in mind: the university undergraduate in geology, physical geography or earth and environmental sciences at a basic rather than a research level, and the lay person interested in understanding the landscape of what is one of the most spectacular and exciting regions of Europe. We do not use within-text references, but at the end of chapter 1 we provide suggestions for further reading.

The book is divided into two parts. In the first part we give a broad outline of the main themes of the regional geology and geomorphology. The emphasis is particularly on the geology and geomorphology of the Neogene sedimentary basins, set within the context of the regional structural geology. The second part is devoted to a series of regionally-based field excursions and site descriptions focusing on key locations, each illustrating a range of themes.

We have had many years of experience in the area, researching the late Neogene and Quaternary geology and the geomorphology, and have published extensively on the area in the research literature. During that time we have benefited from many discussions in the field with numerous colleagues, particularly: Roy Alexander, Astrid Blum,

Juan Braga, Pat Brenchley, Jose Calaforra, Adolfo Calvo, Trevor Elliott, Tony Garcia, Martin Geach, Jose Goy, Jim Griffiths, Andy Hart, Peter Haughton, Suzanne Hunter (née Miller), Mike Kelly, Luna Leopold, Jim Marshall, Jose Martin, Juan Puigdefabregas, Pablo Silva, Alberto Sole-Benet, Diane Spivey, Martin Stokes, Cesar Viseras, Steve Wells, Elizabeth Whitfield (née Maher), Kate Willshaw, and Cari Zazo. We also wish to thank Sandra Mather (no relation!) for her expert cartography; she drew the maps and diagrams. We also thank staff at Dunedin Academic Press for their guidance and patience. Finally AMH wishes to thank his wife Karina for her support, patience and forbearance during the writing of this book.

# Part I

# Main themes in the geology and geomorphology of Almeria

# Chapter 1

# Introduction – highlights of the geology and geomorphology of Almeria

Almeria province is situated at the southeast 'corner' of the Iberian peninsula, not far north of the southern boundary of the European tectonic plate, and its geology has been profoundly influenced by interaction between the European and African tectonic plates. From Mesozoic to mid-Tertiary times that boundary was a collisional one causing compression to the southern margin of the European plate. That compression caused the thrusting northward of a series of nappes onto the European foreland. They originated deep in the crust within the collision zone and comprise dominantly metamorphic rocks, which themselves were formed mostly during the Palaeozoic (for geological timescale *see* Appendix 1). Today they form the main mountain areas of the internal zone of the Betic Cordillera (that is the zone in which Almeria lies: *see* Fig. 1.1), and the basement of the Almeria region. Three distinct nappe complexes have been recognized. The lowest is the Nevado–Filabride complex, though there is some discussion as to whether this is a true nappe transported from the south rather than simply the underlying basement of the region, thrust forward to some extent. Today the Nevado–Filabride nappe forms the mountains of the Sierra Nevada and the Sierra de los Filabres (*see* Fig. 1.2). Overlying this nappe and clearly thrust from the south, probably riding over a base of Triassic marls, is the Alpujarride nappe. This today forms a string of separate mountain ranges south of the Sierras Nevada and de los Filabres. These are, from west to east, the Sierras de Gador, Alhamilla, Cabrera and several Sierras bounding the north of the region, north of the Vera basin (Fig. 1.2). The third nappe, the Malaguide nappe, is important west of our region, but within our region forms a few patchy outcrops resting on Alpujarride rocks in the mountains to the north of the Vera basin. The basement rocks will be described in chapter 2, but their importance for the Neogene and Quaternary geology lies in their role as provenance markers within the

3

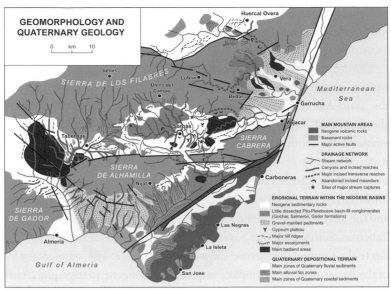

**1.4** Geomorphology and Quaternary geology of the Almeria region. Note the contrasts between Quaternary erosional and depositional terrain.

what is now the Alboran Sea, and was tectonically transported into the area along the Carboneras strike-slip fault system (Fig. 1.2). These rocks now form the low mountains of the Cabo de Gata bounding the Almeria basin to the southeast (Fig. 1.4).

The Neogene to Quaternary sedimentary fill of the basins records the Neogene to Quaternary history of the region. Initially during the Serravallian (Appendix 2) sedimentation was under terrestrial conditions, then for the majority of the Neogene, under marine conditions. During the late Messinian (Appendix 2) the connection between the Mediterranean Sea and the Atlantic Ocean was severed. The Mediterranean Sea became desiccated through evaporation (during the so-called Messinian salinity crisis); sea levels fell, exposing the area to subaerial erosion. When the Mediterranean finally became linked to the Atlantic Ocean, sea levels rose again. During the oscillating sea-level rise pockets of highly saline water became trapped within the more land-locked parts of the basins, causing the precipitation of evaporitic gypsum. By the early Pliocene (Appendix 2) fully marine conditions had been established in the more low-lying parts of the basins; however, this marine phase did not last long and by the late Pliocene and into the Quaternary sedimentation under terrestrial conditions resumed in most areas. We deal with the details of the Neogene sequence in chapter 3 and the Plio-Quaternary switch to terrestrial conditions in chapter 4.

The Neogene to Quaternary sedimentary sequences record not only the effects of Mediterranean-wide controls, but also those of much more local or regional controls. Disregarding the complexities associated with the Messinian salinity crisis (Mediterranean-wide), the overall trend from the Tortonian (*see* Appendix 2) conditions of moderate marine depth to the Messinian to Pliocene shallow marine conditions to the Plio-Quaternary terrestrial conditions reflects the overall 'post-orogenic' uplift of the region. The amount of this epeirogenic uplift varied between mountain ranges and basins and from basin to basin (*see* chapter 4). In a classic study Juan Braga and his colleagues mapped the early Pliocene coastal deposits, recording their present elevations, to give an estimate of the spatial variations in the relative amount of post-early Pliocene uplift within the region. We have elaborated his original data and present these results in chapter 4.

Differential epeirogenic uplift between the basins has led to a sequence of differential marine retreat from the basins, with retreat earliest from the Tabernas, with a Pliocene fan delta formed in the Rioja corridor at the basin outlet. Second came the Sorbas basin, with Pliocene seas rapidly retreating to fan deltas at the margins of the Almeria basin to the south. There were fan deltas within the Vera basin, marine retreat coming only at the end of the Pliocene. Within the Almeria basin, marine retreat from the north of the basin occurred in the late Pliocene, but in the south of the basin a series of offlapping Pleistocene shorelines mark the later marine retreat there.

The same patterns of uplift that brought about a switch from marine to widespread terrestrial deposition later caused a widespread switch from terrestrial deposition to erosion by triggering incision of the emerging drainage network (Fig. 1.4), to form the modern erosion-dominated landscape and the array of landscapes illustrated in Figure 1.5. Other factors, of course, were at work. As well as differential epeirogenic uplift, the region is still tectonically active with deformation continuing on many of the major faults, not only on those that are exposed but on faults in the basement underlying the sedimentary basins, causing deformation of the overlying Neogene to Quaternary sediments. Global climatic changes during the Quaternary caused

**1.5** Almeria landscapes. **A**: Google Earth image – Compare with Figures 1.2, 1.3, 1.4. **B**: Mountain landscapes: Sierra de los Filabres. Note the relatively even summit areas, and the deep valley dissection. **C**: A dissecting sedimentary basin environment: the western part of the Tabernas basin, characterized by deeply entrenched stream channels and badland development on the hillslopes. **D**: An aggrading sedimentary basin environment: the eastern part of the Tabernas basin, characterized by coalescent alluvial fans.

changes in the erosion/sedimentation regime, with periods of sediment excess during the global glacials (though there was no actual glaciation here) alternating with periods of erosional/incisional dominance during the interglacials. The same climatic oscillations caused variations in sea level (sea-level fall during the glacials and highstands as now during the interglacials), particularly affecting the coastal areas. Finally, local changes brought about by re-organization of the drainage network through river capture have had profound effects, not only on water and sediment routing, but also on incisional/aggradational regimes. This is an ideal region to study the interplay between these factors and their influence on landform development. We deal in detail with these themes in chapter 4.

Quaternary climates were predominantly dry. Today the region is the driest in Europe with mean annual precipitation less than 250mm. This, together with the mild winters and hot summers ( July daily maximum temperatures average almost 30°C), results in a truly semi-arid climatic regime. Indeed, the combination of bare erosional landscape and the hot, dry climate has led to the Tabernas area being described as the 'Desierto de Tabernas'. During the Quaternary, aridity had profound effects on the geomorphic processes – effects that continue today. Rainfall, of course, does occur, occasionally as prolonged downpours in storms characterized by high rainfall intensity. Such rainfall, falling on soils with only low infiltration capacity and only a thin vegetation cover (in many cases the vegetation itself having been degraded by many centuries of human activity), causes high rates of runoff and high rates of erosion. In extreme cases of steep slopes on weak bedrock this results in spectacular 'badland' landscapes. With little groundwater recharge, the river systems tend to be dry most of the time, carrying water (and sediment) only during flash-flood conditions. The river channels are characteristic gravel-dominated 'dryland' channels, some showing well-developed braided patterns (albeit dry for most of the time).

Another important aspect of aridity affects soils and pedogenic processes. With little soil moisture availability, pedogenic processes are slow. However, on stable sites over the longer timescales of the late Quaternary the absence of leaching allowed the buildup of various compounds in the soil profile. The high soil temperature, the high pH and the oxygenated nature of the soil allow the buildup of ferric iron oxides, giving the older soils a red colour (the older the soil, the deeper red the colour). Below such a horizon, and because of incomplete leaching and high pH, calcium carbonate builds up at depth in the soil profile (again the amount of carbonate reflects the soil age). On exposure to the atmosphere, possibly through erosion of the overlying horizons, the carbonate becomes indurated to form a pedogenic calcrete. This band of 'rock' may be up to about 1m thick and form an erosion-resistant caprock to many features such as river terraces and alluvial fan surfaces. Also, where limited leaching has been possible, carbonate accumulation has taken place on exposure of the interface between superficial deposits and bedrock, to form a groundwater calcrete.

The region as a whole presents a superb range of 'dryland' features and we explore these in more detail in chapter 5.

## Further reading

### Basic background material

If you need basic background material in geology or geomorphology we recommend the following two books, both published by Dunedin Press. They both deal with the general principles of the subject concerned, assuming a general interest in the subject, but little prior knowledge.

> Graham Park, 2006 (2nd Edition: 2012). *Introducing Geology, a Guide to the World of Rocks.*
>
> Adrian Harvey, 2012. *Introducing Geomorphology, a Guide to Landforms and Processes.*

### More specialist material

These books relate specifically to the geology and geomorphology of Spain, including at least some material relating to Almeria. They assume at least some background knowledge in geology and/or geomorphology.

> Wes Gibbons and Teresa Moreno (Editors), 2002. *The Geology of Spain.* The Geological Society of London. (A blockbuster! Arranged chronologically, covering every part of the geological evolution of Spain – assumes considerable background knowledge in geology.)
>
> Francisco Gutierrez and Mateo Gutierrez (Editors), 2014. *Landscapes and Landforms of Spain.* Springer, Dordrecht. (Assumes some background knowledge: includes three chapters specifically related to Almeria.)

Mather A.E., Martin, J.M., Harvey, A.M. and Braga, J.C., 2001. *A Field Guide to the Neogene Sedimentary Basins of the Almeria Province, South-East Spain*. Blackwell Science, Oxford. (A very specific work aimed at the specialist sedimentologist and geomorphologist: currently out of print, but available as an E-book from Wiley.)

John Lewin, Mark Macklin and Jamie Woodward (Editors), 1994. *Mediterranean Quaternary River Environments*. Balkema, Rotterdam. (Includes several papers of direct relevance to Almeria: assumes background in geomorphology and the Quaternary.)

The journal *Geomorphology* published by Elsevier, Amsterdam, and aimed at the academic and professional geomorphologist, has produced two Special Issues on the geomorphology of Spain – both include material relevant to the Almeria region:

Mather, A.E. and Stokes, M. (Editors), 2003. Long-term landscape development in Southern Spain. *Geomorphology* Volume 50, Issues 1–3, pp. 1–291.

Gutierrez, F., Harvey, A.M., Cendrero, A, Gaecia-Ruiz, J.M. and Silva, P.G. (Editors), 2013. Geomorphology in Spain: Special Issue in honour of Prof. Mateo Gutierrez. *Geomorphology* Volume 196, pp. 1–279.

# Chapter 3

# Neogene geology

## 3.1 The volcanics

One of the most intriguing (and delightful) parts of Almeria is the Cabo de Gata area of volcanic rocks (*see* chapter 10, Excursion 5). These rocks dominate the terrain southeast of the Carboneras fault. The rocks themselves originated as lavas and pyroclastic rocks extruded to the southwest of their present position, into what is now the area of the Alboran Sea, and have since been brought into their modern position by movement along the Carboneras fault system (*see* Figs 1.2, 1.3). They also occur as slivers between the various arms of the Carboneras fault system itself. The rocks are chemically intermediate rocks of the calc-alkaline type (*see* below), indicating that extrusion was through continental crust. They include two groups; the older group was extruded in Serravallian to Tortonian time, the younger group during mid-late Tortonian time (*see* Appendix 2).

The origin of these rocks has been a matter of considerable debate. In plate-tectonics terms, they are not related to subduction. There is no subduction zone below the Betic Cordillera; if there ever had been a subduction zone it had ceased to function by the end of the Cretaceous (*see* Appendix 1). Current thinking is that from the Mesozoic to the Early Cenozoic (*see* Appendix 1) the collision between the African and European plates created a mountain range in the area of what is now the Alboran Sea. From this compressional zone, nappes were thrust north onto the European/Iberian foreland to form the Betic structures (*see* previous chapter) and their mirror-image equivalents were thrust south onto the African foreland to form the Rif ranges in Morocco, in a spatial pattern almost reflecting that of the Betic ranges in Spain (*see* Fig. 1.1). During early-mid Tertiary time the compressional regime relaxed and the root of that 'Alboran' mountain chain collapsed into the upper mantle. This resulted in the development of the Alboran Sea basin during Serravallian–Tortonian time, and the initiation of volcanic activity as partial melting of the former mountain core took place. From the late Tortonian onwards, that tectonic regime was replaced by that of the Trans-Alboran shear zone (*see* previous chapter), developing the deep-seated left lateral strike-slip fault zones. One local result was the translocation of the Cabo de Gata volcanic zone into its present position along the Carboneras fault system. Another result was the initiation of other small late Neogene volcanic centres northeast of the Alboran Sea through our region (*see* below) as far north as the Cartagena area.

The rocks of the Cabo de Gata zone are a series of intermediate calc-alkaline lavas ranging from andesites and dacites to more acid rhyolites, but including a significant proportion of pyroclastic rocks. These rock types dominate the older volcanic group. Grey to buff lavas with lath-like phenocrysts of dark brown hornblende (*see* Fig. 3.1) are common. The pyroclastic rocks are perhaps best seen on coastal exposures south of San Jose (for locations *see* chapter 10). The rocks were extruded below sea level. Tortonian marine sediments are incorporated in the volcanic rocks, especially near the junction between the older and younger volcanic rocks. The older rocks are overlain in part by the younger volcanics, which form

**3.1** Cabo de Gata volcanic rock – hand-specimen scale: hornblende-rich dacite/andesite: stream-bed pebble collected from north of Carboneras.

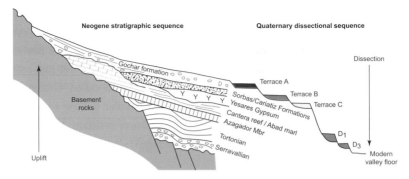

**3.3** Schematic model of the Neogene aggradational sequence in Almeria (left), followed by the dissection sequence of the Quaternary (right). Based primarily on the sequence of the Sorbas basin.

(*see* above), together with the early uplift of the Filabride zone. This combination of extension and uplift created accommodation space on the southern margin of the Filabres, the first phase in the development of the Tabernas and Sorbas basins, and provided a major source of sediment from the uplifting Sierra de los Filabres. The initial Serravallian fill of the incipient Tabernas basin is composed of coarse terrestrially-derived conglomerates, locally including enormous blocks up to 2m in diameter (*see* Fig. 3.4A). The clast content is dominated by garnet-mica schist (*see* Fig. 2.1B). This lithology could only have been derived from the proto-Filabres, presumably from the Mulhacen nappe (*see* chapter 2), prior to the local unroofing of the Filabres, exposing the underlying rocks of the Tahal nappe. Today these Serravallian conglomerates crop out in the south part of the Tabernas basin, forming the anticlinal Serrata del Marchante (*see* Fig. 1.2). Exposed in the Rambla Sierra, southwest of Tabernas village, the transition can be observed from these Serravallian terrestrial debris-flow conglomerates into marine conglomerates and then into the deeper water marine turbidite-dominated sequence of the Tortonian (*see* chapter 7, Excursion 2). The Serravallian conglomerates also crop out on the northern margins of the Tabernas basin, and in the far northwest and southeast of the Sorbas basin.

### 3.2.2 The Tortonian: a turbidite-dominated marine sequence

Rapid subsidence during the (early?) Tortonian led to the development of a marine basin extending W–E through the modern Tabernas, Sorbas and Vera basins and south into the Almeria basin, bounded to the north

**3.4** Serravallian to Tortonian. **A**: Serravallian megaconglomerate – south of Tabernas (large block is more than 2m in diameter). **B**: Tortonian 'Bouma-type' turbidite sequence, near Penas Negras, Sorbas basin (hammer for scale). **C**: Calcareous sandstones of the Azagador Member – northern margin of the Sorbas basin (photo shows about 5m of section).

by the Sierra de los Filabres. Tortonian marine sediments are also present in patches high on the Sierre da Alhamilla. Although this suggests that marine conditions were continuous across the Sierra de Alhamilla, it is likely that the highest parts of this massif and of the Sierra Cabrera were offshore islands.

The Tortonian rocks themselves, best seen in the Tabernas basin (*see* chapter 7, Excursion 2), are dominated by sand/mud turbidites (*see* Fig. 3.4B), fed from the Filabres into a series of what were then submarine fans. In the basin centre are thick grey marls. There are occasional much thicker sandstone units and some conglomerate-filled channel systems that can be traced for several kilometres. One of the most interesting beds is a dark schist-rich debris-flow conglomerate that is extensive throughout the centre of the Tabernas basin, the 'Gordo megabed', which has been interpreted as a seismite. It is the largest and most extensive of several such beds within the Tortonian of the Tabernas basin. Further evidence of ongoing tectonic activity inducing instability in the within-basin sediment pile is the presence of numerous slump folds. Away from the Tabernas basin the Tortonian sequence tends towards sandier turbidites (*see* chapter 6, Excursion 1).

### 3.2.3 Tortonian–Messinian transition: deformation followed by marine transgression and the initiation of a shallow marine sequence

During late Tortonian time renewed tectonic activity, related to the development of the strike-slip fault system, caused folding and uplift of the Tortonian sediments. The overlying Azagador Member of the Turre Formation rests on the Tortonian sediments with a marked angular unconformity. At that time the basins began to emerge as partially distinct from one another, with their limits defined by the uplifting basement units or by the strike-slip fault system itself. The Azagador Member is a resistant fossiliferous sandy limestone (Fig. 3.4C) whose fossil content indicates a temperate climatic environment in late Tortonian time. Today the Azagador Member forms outfacing escarpments that define the inner (Messinian) sedimentary basins (*see* chapter 6, Excursion 1). The Azagador Member flanks both northern and southern margins of the eastern part of the Tabernas basin, both margins of the Sorbas basin, the southern margin of the Vera basin and the northern margins of the Almeria basin.

### 3.2.4 Early Messinian: basin marginal coral reefs, basin centre marls

The Azagador Member was succeeded in early Messinian time by the Cantera and Abad Members of the Turre Formation. Our colleagues from the University of Granada, Jose Martin and Juan Braga, have studied the fossil content of these rocks and demonstrated that the climatic conditions had warmed from temperate to tropical conditions by the Early Messinian. The Cantera Member is a coral-reef limestone (Fig. 3.5A,B), whose morphology is beautifully preserved in a number of locations, for example near Nijar on the northern margins of the Almeria basin (*see* chapter 10, Excursion 5) and near Cariatiz on the northern margins of the Sorbas basin (*see* chapter 6, Excursion 1). It also rests on basement crossing the divide between the Sorbas and Almeria basins at Cantona hill, a high point in the structural low between the Sierras Alhamilla and Cabrera. This link

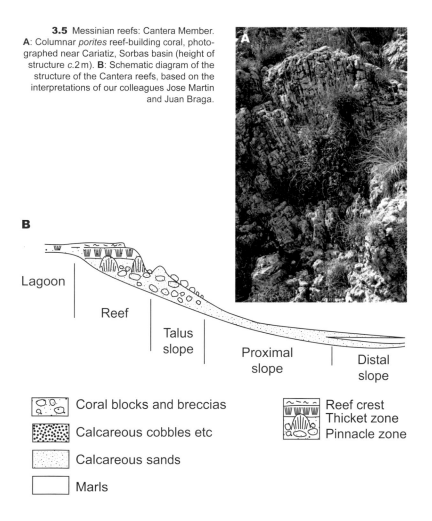

**3.5** Messinian reefs: Cantera Member. **A**: Columnar *porites* reef-building coral, photographed near Cariatiz, Sorbas basin (height of structure *c.*2 m). **B**: Schematic diagram of the structure of the Cantera reefs, based on the interpretations of our colleagues Jose Martin and Juan Braga.

**3.7** After the Messinian salinity crisis. **A**: Clean cross-bedded sandstones of the Sorbas Member, W of Sorbas (photo shows *c.*2 m of section). **B**: Stromatolite dome, Sorbas Member, near Moras (dome is *c.*2 m high).

### 3.2.7 Latest Messinian–Pliocene: culminating in marine retreat

Up to this point there was some commonality in the Neogene sedimentary sequences among the four sedimentary basins, but from this timepoint on the sequences become increasingly disparate. The differences are related to differential uplift and topographic emergence as marine conditions are progressively replaced by terrestrial conditions from the latest Messinian through the Pliocene and into the Quaternary (*see* chapter 4). In the bulk of the Tabernas basin (except within the downfaulted westernmost part of the basin, the Andarax valley) there is no evidence of marine deposition following the patches of gypsum. The Tabernas basin is the basin with the greatest amount of post-Pliocene uplift (*see* chapter 4), so we can assume that it had become emergent by the late Messinian.

Within the Sorbas basin the shoreline retreated south and the shallow marine sediments of the Sorbas Member are succeeded by the terrestrial sediments of the Cariatiz Formation. These include low energy, red to buff coastal plain silts of the Zorreras Member in the basin centre but sandstones and conglomerates of the Moras Member further north (*see* chapter 6, Excursion 1). The latter have been interpreted as laid down in small alluvial fans fed from the Sierra de los Filabres. Clastic sedimentation during deposition of the Zorreras Member was interrupted by three phases of lagoonal conditions during which thin carbonates were deposited. These carbonates act as basinwide marker bands (Fig. 3.8A) and the spatially variable thicknesses of the intervening reddish silts are indicative of ongoing tectonic deformation within the Sorbas basin. The top of the Cariatiz Formation, of lower Pliocene age, is marked by a marine band of fossiliferous yellow sands (Fig. 3.8B) in the basin centre and a fossiliferous sandy pebble bed in the north of the basin.

**3.8** End Miocene–Pliocene. **A**: Zorreras Member, white lacustrine band – Sorbas. **B**: Final-Zorreras Member, soft marine sandstone band – North of Sorbas. (Both photos show sections of c.1 m.)

**3.9** Espiritu Santo Formation, fan-delta sediments – Vera basin. (Photo by kind permission of Dr. Martin Stokes.)

Both the Vera and Almeria basins remained marine throughout the latest Messinian and into the Pliocene. In both basins the shallow marine sandstones of the lower Pliocene Cuevas Formation, fine yellow bioturbated sandstones, are succeeded by coarser, wave-influenced sandstones. These are overlain locally by the Espiritu Santo Formation. This pebbly quartz-conglomerate was deposited in fan-delta environments (Fig. 3.9) fed by fluvial input into the eastern part of the Vera basin and the northern part of the Almeria basin. In addition, in the throat between the northwest corner of the Almeria basin and the western/distal end of the Tabernas basin is another large body of Pliocene fan-delta conglomerates, the Abrioja Formation. These sediments were fed by the Andarax river system from the western end of the Sierra de los Filabres.

### 3.2.8 Late Pliocene– Early Pleistocene: marine retreat and replacement by terrestrial conditions

The sediment sequences of the late Messinian–Pliocene clearly represent the effects of regional uplift. By the mid-late Pliocene the sea had already retreated from the Tabernas basin, except from the throat of the Rioja corridor between the Tabernas and Almeria basins. It had also retreated entirely from the Sorbas basin, and was retreating from the Vera and Almeria basins. The uplift of the mountain blocks generated a new wave of sediment production that resulted in the deposition of coarse alluvial-fan and fluvial conglomerates within the sedimentary basins. The Pliocene Gador Formation in the Andarax valley to the west of the Tabernas basin comprises coarse fluvial conglomerates, fed from the western end of the Sierra de los Filabres, and feeding into the fan delta in the Rioja corridor. As further marine regression took place fluvial conglomerates prograded downstream along the Andarax valley into the Almeria area.

The Pliocene to early Pleistocene Gochar Formation is the equivalent in the Sorbas basin. It comprises alluvial-fan conglomerates (Fig. 3.10) fed into the margins of the basin primarily from the Sierra de los Filabres, but also in the south of the basin from the Sierra de Alhamilla. These fans fed an axial river system that flowed south out of the basin across the structural low between the Sierras Alhamilla and Cabrera into the northern tip of the Almeria basin (Fig. 3.11). This fluvial system was the forerunner of the dominant Quaternary drainage system of the Sorbas basin, the Aguas/Feos system (see chapter 4). At the margin of the Almeria basin it fed a fan-delta

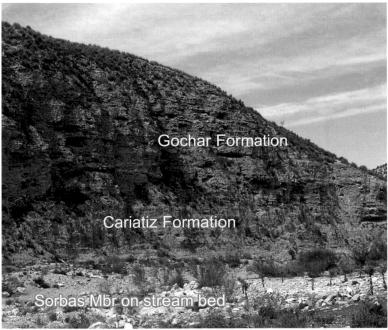

**3.10** Gochar Formation, final basin-filling phase (Sorbas basin). Gochar alluvial-fan conglomerates – Sorbas basin. Note: at the base the relatively fine fluvial sands and gravels of the Cariatiz Formation (Moras Member) cut into and overlain by Gochar Formation coarser braided-channel gravels – note palaeosol horizons.

# Chapter 4

# The Quaternary: uplift, emergence and landform development

## 4.1 Uplift and marine withdrawal: the Pleistocene shoreline sequence

As we have seen in the previous chapter, differential epeirogenic uplift amongst the four basins caused a switch from marine to terrestrial environments. The amounts of post lower-Pliocene uplift are illustrated in Figure 4.1, with greatest amounts in the Tabernas, followed by the Sorbas then the Vera basins, and least in the Almeria basin. The timing of marine withdrawal reflects these patterns. By the early (?) Pliocene, marine sedimentation at the outlet of the Tabernas basin was confined to the fan delta in the lower Andarax valley in the throat of the Rioja corridor. Marine retreat took place in the Sorbas basin after deposition of the short-lived early Pliocene Zorreras/Cariatiz marine band (*see* chapter 3). Marine conditions persisted into the Pliocene in both the Vera and Almeria basins.

In the Vera basin marine retreat occurred after the fan-delta phase of the Pliocene Espiritu Santo Formation (*see* chapter 3) was replaced by coalescent alluvial fans of the early Pleistocene Salmeron Formation. These fans extended towards the basin centre, which, according to our colleague Martin Stokes, was by then a zone of interior drainage. In the north of the Almeria basin small Pliocene fan deltas, similar to the Espiritu Santo Formation in the Vera basin, were followed by an early Pleistocene fan delta (the Polopos Formation) of the proto-Aguas/Feos system (*see* chapter 3). This fan delta marks the final marine phase in that part of the basin. Further west, in what is now a relatively uplifted block of terrain between Nijar and the Andarax valley, a series of early and mid-Pleistocene shoreline sediments can be traced from the basin margin offlapping towards the basin centre. This indicates mid-late Pleistocene shoreline retreat from

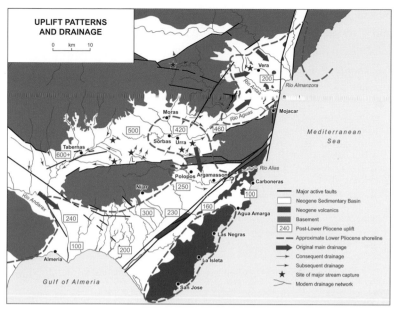

**4.1** Pliocene shorelines and post-Pliocene uplift patterns (uplift patterns modified from the work of Juan Braga). Boxed green figures show the modern elevations of lower Pliocene shoreline sediments, and therefore give an indication of the *relative* amounts of post-Pliocene uplift. Note the differing degrees of uplift between the four sedimentary basins. Note also the main drainage outlets – the Vera basin system; the Sorbas basin drainage into the northern part of the Almeria basin; and the Tabernas/Andarax system.

the uplifting Sierra de Alhamilla. There is a similar situation in the Dalias lowland, the near coastal area west of Almeria, south of the Sierra de Gador.

During the mid-late Pleistocene, sea levels were affected not only by regional tectonics, but also by global eustatic sea levels related to the global glacial sequence, with high sea levels (as now) during global interglacials and low sea levels during global glacials (*see* Appendix 3). Remnants of mid-late Pleistocene interglacial cemented beach sediments are preserved at several localities along the coast. There are a few scattered remnants of what appears to be an early raised beach related to either the penultimate or an earlier interglacial (MIS Stage 7 or 9: *see* Appendix 3) about 20m above modern sea level. There are more extensive remnants of the raised beach related to the last interglacial, (MIS Stage 5, or using older Mediterranean terminology, Tyrrhenian 3; *see* Appendix 3), containing the index gastropod fossil, *Strombus bubonius*, characteristic of (sub)tropical environments, but not present in the Mediterranean today. These sediments occur up to about 8m above modern sea level in a number of localities in our area.

Several have been studied in detail and dated by our colleagues Jose Goy and Caridad Zazo. These include the Dalias shore, Almeria Bay, several localities in the Cabo de Gata zone and between Carboneras and Mojacar (*see* chapter 9, Excursion 4: *see* also Fig. 4.2).

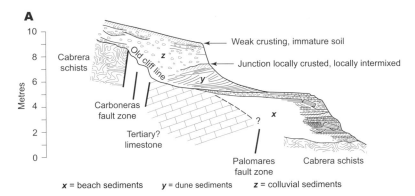

**A**

Weak crusting, immature soil

Junction locally crusted, locally intermixed

Cabrera schists

Carboneras fault zone

Tertiary? limestone

Palomares fault zone

Cabrera schists

**x** = beach sediments      **y** = dune sediments      **z** = colluvial sediments

**4.2** Pleistocene last interglacial coastal deposits. **A**: Summary of the Pleistocene shoreline stratigraphy at Macenas; *x*: beach sediments, *y*: dunes, *z*: colluvium (see chapter 9, Excursion 4). **B**: Photo of the sequence shown in Fig. 4.2A. **C**: Fossil assemblage from last interglacial raised beach sediments at Macenas: note the gastropod *Strombus* (arrowed) – albeit a small one!).

## 4.2 The initial drainage: incision

Now we must return to the Quaternary evolution of the basins themselves. As the late Pliocene–early Pleistocene marine withdrawal from the basins was time-transgressive, so too was the initial phase of terrestrial sedimentation. The Gador Formation (Tabernas basin) almost certainly dates from the Pliocene, the Gochar Formation (Sorbas basin) from the Plio-Pleistocene, but the Salmeron Formation (Vera basin) more likely dates, at the earliest, from the latest Pliocene. The equivalent conglomerates at the northern end of the Almeria basin almost certainly date from the early Pleistocene. The landforms associated with this phase of fluvial gravel deposition were large coalescent alluvial fans grading into axial river systems. Any remnants of the original surface topography have since been eroded from the Tabernas basin, but are preserved as gently sloping cemented gravel surfaces in the northern parts of the Sorbas and Vera basins and to a lesser extent in the northern portion of the Almeria basin.

The rivers initiated on this landscape (the original consequent drainage), were related to the gradients generated by the uplift patterns and by the location of the main structural blocks and areas of ongoing tectonic deformation (Figs 4.1, 4.3). These rivers were the forerunners of the

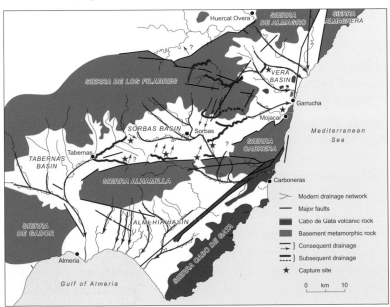

**4.3** Map of drainage evolution showing original consequent drainage directions, modified by subsequent stream development, involving numerous stream captures.

modern drainage. In the west, during deposition of the Pliocene Gador Formation, the proto-Andarax was fed by the uplifted Filabres and the eastern Alpujarras and the Tabernas basin from the east. The main stem exited to the south, then as now, through the Rioja corridor between the Sierras de Gador and Alhamilla into the Almeria basin. In the Sorbas basin the main drainage during sedimentation of the Gochar Formation was centripetal into the basin from the surrounding Sierras de los Filabres and Alhamilla (*see* Fig. 3.9B). It exited the basin to the south into the north of the Almeria basin through the structural low between the Sierras de Alhamilla and Cabrera. We call this drainage the proto-Aguas/Feos. Within the Almeria basin it initially fed a fan delta in the north of the basin (depositing sediments of the Polopos Formation, *see* above). With marine retreat, the river then probably followed the structurally weak zone roughly where the Cabrera southern boundary fault zone converges with the Carboneras fault zone, in other words roughly along the course of the modern Rio Alias. The main tributary within the Almeria basin was a right-bank tributary, the proto-Alias, fed from the Sierra de Alhamilla. The Vera basin was initially an area of internal drainage, but when the basin became open to the sea it was drained by several small streams, including the proto-Antas, fed from the eastern end of the Filabres. The proto-Almanzora, now the main river, seems to have been simply a stream draining the Sierra Almagra. The Huercal Overa basin, northwest of the Sierra Almagra, now drained by the Almanzora, was also an area of internal drainage, apparently only later to be 'captured' by the Almanzora.

During the early-mid Pleistocene the river systems began to incise through the Plio-Pleistocene conglomerates into the underlying Neogene basin-fill sedimentary rocks. In doing so, the weaker rocks were exploited by the development of subsequent streams, leading to stream capture and the reorganization of the drainage pattern towards the modern drainage systems (*see* below, section 4.3). During incision, superb exposures of the Neogene rocks were created, for the benefit of geologists! Perhaps, more importantly, the basins, which during the Neogene had been sediment sinks, now became dominantly erosional (Fig. 3.3, right-hand side) with a net export of sediment. The erosional landscapes create the spectacular scenery of the region. Our interpretation of the regional geomorphology, in terms of dominantly erosional or depositional landscapes, is illustrated in Figure 1.4 (for erosional landscapes *see* also Fig. 4.4).

## 4.3 Quaternary erosional landscapes

River incision is driven by stream power; that is, by river discharge and gradient. It is impeded by resistant substrate. It is not surprising that the Tabernas basin is the most deeply dissected basin; it has the steepest tectonically-induced gradients, between the most elevated basin and the low in the Rioja corridor. The basin-fill comprises weak Tortonian marls and turbidites. The least dissected basin is the Almeria basin, which has the least uplift and the lowest tectonically-induced gradients.

Where the incising rivers encountered resistant bedrock, deep, narrow canyons were formed (Fig. 4.4A). Examples are the canyon south of Tabernas village, cut into resistant Serravallian conglomerates (*see* chapter 7, Excursion 2), and the Aguas canyon near Los Molinos cut through the gypsum (*see* chapter 6, Excursion 1). It is odd, perhaps, that the gypsum, a soft mineral, actually forms a resistant rock in this area. This is partly because of the bond between the gypsum crystals, accentuated by the dry climatic conditions. In less resistant bedrock the incised valleys are wider, and the rivers tend to migrate laterally as well as to incise vertically. This has led to the development of some spectacular incised or ingrown valley meanders, some of which have 'cut off' leaving abandoned valley meander loops perched above the modern valley floor. Sorbas village sits on an 'island in the sky' (*see* chapter 6, Excursion 1) created by a complex cutoff at the junction between two incised valleys. Such tortuous incised meandering valleys and associated complex cutoffs tend to occur especially where neotectonic activity has modified stream gradients, either by tilting or by faulting. Examples are in the Moras area of the Sorbas basin (*see* chapter 6, Excursion 1), and in the Almeria basin where the Rio Alias crosses the Carboneras fault zone. That locality has been studied by Elizabeth Whitfield where she demonstrated the repeated effects of faulting on river incision and on the development of cutoffs (*see* chapter 9, Excursion 4).

One important aspect of drainage network development that has fundamental implications for the functioning of the landscape is river capture. This is where one river incises more rapidly and by headwards erosion intersects the course of another river, so diverting the flow (and sediment) in a new direction. This is usually brought about where the aggressive stream (a 'subsequent' stream) flows preferentially along a weaker rock but intersects the course of an earlier stream that may have been part of the original drainage (a 'consequent' stream). In our area there are many examples of river/

**4.4** Representative erosional landscapes. **A**: Incised meandering canyon, cut into Serravallian conglomerates – northern margin of the Tabernas basin. **B**: Badland landscape, Tabernas basin. **C**: Escarpment in Neogene rocks. View west from Penas Negras, southern margin of the Sorbas basin. Note: the main scarp-formers here are the Azagador Mbr sandstones (mid-slope) and the Yesares Mbr gypsum(capping the sequence).

stream capture modifying the original drainage directions, but one stands out as having huge implications for the later development of the geomorphology. This is the capture of the original proto-Aguas/Feos drainage (*see* above – chapter 3) by the lower Aguas. The lower Aguas, incising rapidly into the weak Abad marl cut back its headwaters to intersect the alignment of the proto-Aguas/Feos near Los Molinos (*see* chapter 6, Excursion 1). The effects were to divert the flow of water (and sediment) from the original course southwards across the Sorbas basin towards the Almeria basin into a new course eastwards towards the Vera basin (*see* Fig. 4.5). The lower Aguas greatly increased its drainage area and hence its discharge. The Rambla de los Feos was beheaded, leaving only a dimunitive misfit stream following the original course of the proto-Aguas/Feos towards the Alias system in the Almeria basin. Another important effect was the lowering of base level at the capture site at Los Molinos by about 90m. This triggered an incision wave to work its way upstream through the centre of the Sorbas basin, leaving the old valley floor as a terrace perched high above the modern river bed (*see* below).

There are numerous other captures in our area (*see* Fig. 4.3). Drainage into the northwest corner of the Sorbas basin has been captured by Tabernas drainage. The headwater of the Rio Alias, the Rambla Lucainena, has worked its way along the southern margin of the Sorbas basin, capturing what was originally Sorbas basin drainage emanating from the Sierra de Alhamilla. Within the Vera basin the Rambla Balabona used to flow southeast through the broad open valley in which the town of Vera is situated, but has been captured by the Rambla del Carjete, a steep tributary of the Rio Antas, so now flows into the Antas. Within the Sierra Cabrera at Sopalmo is evidence of an intriguing capture, whose origins are still far from certain (*see* chapter 9, Excursion 4). Along the coast the lower Aguas originally flowed northeast towards Garrucha but now flows through a steep course eastwards to the sea at Mojacar. Perhaps this capture is related to incision during lowered sea levels during the Pleistocene. Low Pleistocene sea levels have also been seen as the cause of incision (but without involving any capture) on the lower reaches of the Rio Alias by Elizabeth Whitfield.

Incising river beds provide the local base level for slope evolution. Where incision continues, the canyon walls become undermined and unstable, triggering rockfalls and landslides (for example, at Los Molinos in the Sorbas basin; *see* chapter 6, Excursion 1). On weak rocks, slopes are oversteepened, accelerating slope erosion, locally triggering gully erosion and badland development (Fig. 4.4B). There are extensive badlands cut into Tortonian marls and turbidites in the Tabernas basin (the 'Desierto de Tabernas'), accentuated by the rapid incision of the drainage network. There are also extensive badlands cut into the Abad marl in the Los Molinos area and downstream along the valley sides of the incised Rio Aguas. Other badlands and gully systems occur on Plio-Pleistocene silts in the La Cumbre area in the southeast of the Sorbas basin, deeply incised by headwater streams of the Rio Aguas (*see* chapter 6). There are also badlands on Messinian marls in the Vera basin, where vertical incision is less, so the badlands are dominated by mini-pediment forms rather than by incising gully systems. Badland processes will be discussed in chapter 5.

Away from incising rivers, where the local base level is stable for a while, slope erosion on weak rocks produces a concave slope. Upslope there may be a more resistant rock, forming a caprock escarpment (Fig. 4.4C). The more resistant rocks within the Neogene sedimentary sequences (the thicker sandstone bands within the Tortonian turbidites, the Azagador calcareous sandstones, the reef limestones, the gypsum) all form caprock escarpments. Where local base level remains stationary for a longer period of time, low-angle slopes develop, forming almost flat pediment surfaces (for example, see the El Cautivo area, Tabernas: *see* chapter 7, Excursion 2).

## 4.4 Quaternary depositional landscapes

Within what is primarily a dissectional landscape, controlled in general by the regional tectonic patterns, there are spatially and temporally restricted zones of deposition (*see* Fig. 1.4). The most important, perhaps, are river terrace sediments. These are, of course, spatially restricted to past and present river valleys. River incision was not continuous, but episodic (*see* Fig. 3.3 , right hand side). Between periods dominated by incision there were periods of excess sediment supply to the rivers, during which times the river valleys aggraded. The Quaternary fluctuations in sediment supply appear to be climatically related. Although during the global glacial phases of the Pleistocene there was no glaciation in our area (there were glaciers in the Sierra Nevada), the higher mountains, especially the Sierra de los Filabres, experienced periglacial conditions. Mechanical weathering by frost shattering would have been high, yielding large volumes of sediment to the stream systems. The major rivers aggraded during the Pleistocene glacial phases (*see* below). During the next incision phase these sediments would be abandoned and perched above river level as a river terrace (Fig. 4.5).

The terraces of the Rio Aguas, the most thoroughly studied river system, were first mapped by AMH and colleague Stephen Wells. We identified three main terraces (and labelled them A,B,C – oldest to youngest: *see* Fig. 4.5) that post-date the initial incision of the Gochar depositional surface, but pre-date the capture event at Los Molinos (*see* above; *see* also chapter 6, Excursion 1). At the capture site, Terrace C forms the floor of the abandoned valley. The three terraces can be traced south along the modern Feos valley into the Alias drainage within the Almeria basin. We realized that the degree of soil development on the three terrace surfaces reflects their relative ages. The soils themselves are characteristic of semi-arid region soils, indicating that semi-aridity was also the dominant soil-forming regime throughout the Pleistocene. Below a thin, partly organic, surface layer, the 'B' horizon of the soil tends towards redness, reflecting the presence of ferric iron oxides produced in this oxidizing environment. The thickness of the 'B' horizon increases with the age of the soil, as does the deepness of the red colour (*see* Fig. 4.6A). Because the soils are dry, complete leaching does not occur, but calcium carbonate accumulates within the profile, below the 'B' horizon. Initially this takes the form of carbonate coatings on stones and within cracks in the soil, then builds up to a thick, continuous layer (Fig. 4.6B,C). As with the increasing redness of the B horizon, the amount and status of the carbonate relates to soil age (*see* Fig. 4.6D). Using these soil properties AMH and Stephen Wells applied an age calibration based on other semi-arid regions and estimated the age of the soil on Terrace C (i.e., immediately before the capture at Los Molinos) to be between 50ka and 80ka. Since the original work other researchers have used various modern dating methods to date the sequence more precisely (*see* Fig. 4.6D). In summary: dissection of the Gochar surface appears to have started before 500ka, with overall dissection since then (*see* Fig. 3.3 right-hand side), but with major aggradational phases culminating around 400ka (Terrace A), culminating around 200ka (Terrace B), culminating around 70ka (Terrace C). In the Sorbas basin these all relate to the proto-Aguas/Feos river system. There are younger terraces (we labelled D1, D3 and E): these post-date the capture and appear to date from ~30ka, 15ka,

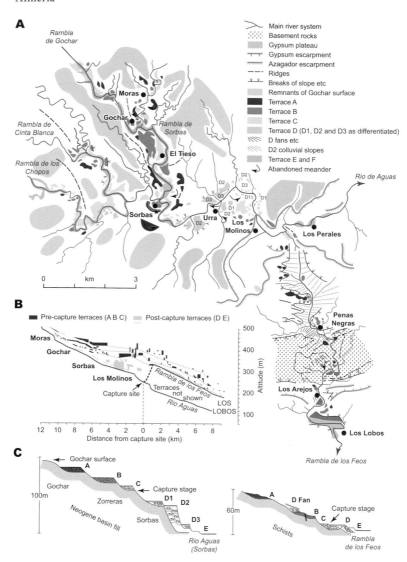

**4.5** Quaternary river terraces of the palaeo-Aguas/Feos river. **A**: Map of the terrace remnants (based on field mapping). Note how all three of the pre-capture terraces (A,B,C) can be traced south through the Feos valley across a sliver of the Sierras into the Alias drainage in the northern part of the Almeria basin. Terraces in the lower part of the Rio Aguas of a similar age to Terrace C in the Sorbas area occur several kilometres downstream of the capture site near Los Molinos, but do not extend upstream into the area of the capture itself. The capture col, southeast of Los Molinos, is floored by Terrace C deposits, so the capture took place during Terrace C times (now dated to c.70ka BP). Following the capture, a massive incision wave worked upstream from the capture site through Sorbas into the headwaters of the drainage. This was followed by further terrace development (Terrace D, locally subdivided into D1, D2, D3 phases, particularly in the Urra area: *see* chapter 6). Along the Feos drainage, the beheaded limb of

the former Aguas/Feos system, there has been little or no incision into the valley floor since Terrace C times. Terrace C along this valley is locally buried by Terrace D. **B**: Profile of the Aguas/Feos system. Note how the pre-capture Terraces (A,B,C above) can be traced south along the Feos Valley, ultimately into the valley of the Rio Alias (*see* Excursion 4, chapter 9). Again, note the lack of incision there since the youngest of the pre-capture terraces (Terrace C). **C**: Valley cross-sections to illustrate the contrasts between the Aguas valley above the capture site with the Feos valley below the capture site. Note that in the Aguas valley, Terrace D is complex and set well below the base of Terrace C, but in the Feos valley Terrace D partially buries Terrace C. **D**: Terraces *A* (right) and *B* (centre) upstream of Sorbas. The terrace deposits rest unconformably on the Pliocene sedimentary sequence.

and the Holocene respectively (Terrace D2 is entirely local in the Urra area near Sorbas: see Excursion 1, chapter 6).

The fact that similar terrace sequences have been identified on the other main rivers of the region, on the Almanzora by AEM and Martin Stokes, and on the Alias by Elizabeth Whitfield, implies that they relate to regional rather than to local controls. The dating of the terraces strongly suggests association of terrace aggradation with excess sediment supply during the glacial phases of the Pleistocene.

One further aspect of soil development is that with time, and especially on exposure, the carbonate layer becomes indurated and resistant to erosion. With age, the lithology of the calcrete becomes more complex, evolving from simple clast coatings to a continuous carbonate horizon. Later brecciation and recementation may take place (Fig. 4.6C) and the surface may be mantled by laminar calcrete. Forming under these conditions it would be referred to as a 'pedogenic calcrete' and may form a caprock over weaker horizons below. Calcrete can also form as a 'groundwater calcrete' where carbonate precipitation occurs in areas of

B

K

Parent material (fan gravels)

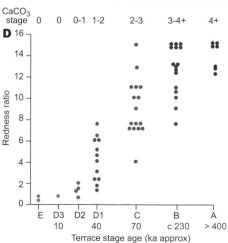

CaCO₃
stage  0    0    0-1  1-2      2-3      3-4+     4+

D

**4.6** Pedogenic calcretes and soils. **A**: Soil profile development on alluvial fan gravels, Tabernas. Note the advanced red colour development of the B horizon, but the relatively immature carbonate accumulation status. This reflects the abundance of pyrite in the parent schists, but the relative scarcity of carbonate-rich rocks in the source area (*see* text). **B**: Calcrete caprock over fan gravels, Nijar. **C**: Brecciated texture of a mature calcrete: Nijar. **D**: Terrace surface soil redness and carbonate status in relation to best estimate of terrace age. Data from the terraces of the Sorbas basin. CaCO₃ status of pedogenic carbonate and calcrete development ranges from 0 (no presence of soil carbonate), through 1 and 2 (minor filaments in cracks and as clast coatings) through 3 (a complete carbonate horizon) to 4 (showing cementation and brecciation). Redness ratio is based on the Munsell colour classification. (The tints shown simply give some indication of the average colour for each terrace sample.) Terraces A–E as identified in the Sorbas basin, with the best current estimates of terrace age in ka BP.

| | | | | | | |
|---|---|---|---|---|---|---|
| E | D3 | D2 | D1 | C | B | A |
| | 10 | | 40 | 70 | c 230 | > 400 |

Terrace stage age (ka approx)

Redness ratio

carbonate-rich percolating water, at the interface between regolith and the underlying unweathered bedrock. David Nash and Roger Smith have identified the layered caprock on the two large early (?) Pleistocene con-glomerate mesas to the northwest of Tabernas village as a combination of pedogenic and groundwater calcretes (*see* chapter 7, Excursion 2). Simi-larly, David Nash and Martin Stokes have identified the calcrete capping mid (?) Pleistocene alluvial fan gravels at Lisbona, in the northwest of the Vera basin, as dominantly a groundwater calcrete.

Aggradation of fluvially-derived sediment may occur not only from free-flowing rivers, but also as a result of ponding. In the west of the Tab-ernas basin a growth fold developed in the underlying Serravallian and Tortonian sedimentary rocks, aligned NW across the basin, probably as a result of movement on a blind fault in the basement underneath (*see* chapter 7, Excursion 2). During the mid-late Pleistocene this uplift caused ponding of the drainage of the Rambla de Tabernas. The result was sedi-mentation of horizontally bedded, fine-grained muds, sometimes lami-nated, sometimes including organic matter. This suggests sedimentation in a palustrine rather than a true lacustrine environment, in other words a swamp rather than a true lake. The 'lake' is in two parts, a lower lake in the centre of the basin and an upper lake upstream of Tabernas, where further ponding was facilitated by small fans constricting the narrow valley. The sediments are up to 20m thick, but thin out downstream and become tec-tonically deformed towards the flank of the anticline. Several years ago during motorway construction, there were excellent temporary exposures of the deformed sediments (*see* Fig 7.7B), but unfortunately these are no longer accessible. There are some dates on the sequence. Luis Delgardo, using the U–Th dating method (*see* Appendix 3) dated rodent bones found at the base of the sequence to ~150ka. This date is now thought to be unreliable; more recent work by Martin Geach now suggests that the initiation of the 'lake' was much later. The top of the sequence (dated by OSL dating by AEM and Martin Geach; *see* Appendix 3) dates from ~20ka. Since that date the Rambla de Tabernas has cut through the barrier caused by the uplift, triggering a recent wave of incision within the basin. Interestingly, that incision wave has so far only worked its way headwards through part of the basin (*see* below).

In addition to river terraces, the other important depositional land-form is the alluvial fan. This is a body of sediment deposited where a steep

The style of alluvial fans reflects the mountain-front setting. Steep mountain fronts along active major faults, especially where the adjacent drainage is incised, do not support alluvial fans (e.g., the Lucainena mountain front of the Sierra de Alhamilla), or only support fans from the largest mountain drainages (e.g., Rambla Sierra fan, Sierra de Alhamilla, Tabernas basin). Otherwise, faulted mountain fronts support fans with little back-filling into the feeder catchments, and apices more or less coincident with the mountain front (e.g., Marchante fans, Tabernas basin). Fans on mountain fronts that are tectonically tilted rather than faulted (e.g., Filabres fans, Tabernas) have very irregular upper outlines with considerable backfilling of fan sediments into the mountain catchments. There is a similar situation where the mountain front is defined by the coast rather than by tectonics (Cabo de Gata fans).

Using the same soil-based approach as for river terraces, several phases of fan sedimentation can be identified, again with the main phases of aggradation probably coincident with the Pleistocene glacial phases, and dissectional phases with interglacials. The Cabo de Gata coastal fans show an interesting relationship between proximal and distal controls. Conventionally, we could expect fan dissection when sea levels (i.e. base level) is low, in other words during Pleistocene glacials. However, the potential base-level signal appears to have been overridden by the sediment signal. Sedimentation phases, as elsewhere in our area, coincide with glacials, when sea levels were low. At those times the fans simply prograded out onto the exposed sea floor. Incision may have been inhibited by low off-shore gradients. During interglacials coastal erosion, as a result of high sea levels, caused profile foreshortening, accentuating fan incision (for more information *see* chapter 10, Excursion 5).

# Chapter 5

# Modern active geomorphic processes

In the previous chapter we dealt with the Quaternary evolution of the area and necessarily with the landforms. In this chapter we focus specifically on modern geomorphic processes. In many ways these processes are typical of dryland or semi-arid environments, but there are some processes, particularly those related to gypsum outcrops, that are unusual and exceptionally well developed in this area.

## 5.1 Weathering (including gypsum karst processes)

In addition to a wide range of 'normal' weathering processes there are two sets of particular processes that are either characteristic of semi-arid regions or are particularly well developed in our area. The first group relates to case hardening. On exposed permeable rock faces rainfall soaks into the surface of the rock, then on desiccation, moisture is drawn back to the rock surface by evaporation, bringing with it dissolved salts from the rock. On evaporation of the moisture, the salts are precipitated in the surface layers of the rock, strengthening and hardening those layers. If this outer 'case-hardened' layer is then fractured, removal of the now weaker material below may take place to produce 'honeycomb' weathering, termed tafoni (Fig. 5.1). There are excellent examples of tafoni developed in some of the volcanic ashes exposed along the Cabo de Gata coast. A particular type of case hardening relates to the transformation of a pedogenic or subsurface carbonate horizon into a pedogenic or groundwater calcrete (described above in chapter 4; *see* Fig. 4.6B).

A very specific case of a similar wetting/drying/precipitation phenomenon relates to exposed gypsum surfaces. The gypsum contains both the hydrated form (gypsum itself), and the anhydrous form (anhydrite). Gypsum has a lower density and higher volume than anhydrite. On wetting of an exposed surface, gypsum is preferentially formed, resulting in an increase in volume. This raises the surface in a 'blister', also

**5.1** Characteristic weathering form: tafoni, developed on volcanic rocks, near Carboneras (section c.5 m).

**5.2** Gypsum karst features: **A**: Tumulus, near Moras. **B**: Doline, gypsum plateau, eastern part of the Sorbas basin. (Approx. feature scales A: c.2 m across; B: < 1 m.)

described as a gypsum tumulus (though of course, totally unrelated to the archaeological features). There are well-developed gypsum blisters at a range of scales on the gypsum plateau in the east of the Sorbas basin (Fig. 5.2A) and other gypsum karst features, such as sinkholes (or dolines) (Fig. 5.2B). A word of warning: the features are delicate. That area is part of the protected landscape of the 'Paraque Natural' – access is restricted. However, similar features are seen on one of the stops included in Excursion 1 (*see* chapter 6). Gypsum is also soluble in rainwater, so rainwater that does enter the ground through fissures dissolves the gypsum, but also erodes the weak marls interbedded within the gypsum. This leads to cave formation, especially through the erosional removal of the marls to form subterranean 'interstratal karst', a term coined by Jose Calaforra, who has studied the Sorbas gypkarst phenomena in detail. The dissolved gypsum may also be precipitated within the caves as stalactites and stalagmites. There are accessible caves near Sorbas – access is achieved through the Sorbas cave club. Gypsum caves are extremely unusual; these caves are well worth a visit.

## 5.2 Slope processes

There are two main forms of important active slope processes. The first is slope failure. Rapid incision by the river systems has created oversteepened, unstable valley-side slopes that may lead to slope failure, generated at the base of the slope. In resistant rocks this might lead to toppling failures (Fig. 5.3A) and rockfalls. In uniformly weak rocks, basal erosion may lead to rotational failures of the slope above. Complex landslides develop where there are interbedded weak and strong rocks, especially if the general dip is into the valley (Fig. 5.3B), such as the features in the Sorbas basin, studied by Jim Griffiths, Andrew Hart and colleagues. Such landslides

**5.3** Slope failures: near Los Molinos. Slope failures triggered by the rapid (post-capture) incision of the Rio Aguas at the base of the slope. The slope itself is composed of relatively strong calcareous sandstones of the Azagador Mbr, dipping fairly steeply northwards towards the river, resting unconformably on Tortonian marly rocks. **A**: Simple toppling failure. **B**: A more complex composite landslide, with the main mass of the Azagador rock moving as a planar slide down the plane of unconformity. Secondary rotational failures occur in the centre of the slide and small toppling failures occur around the margins.

involve planar slides of the resistant bed towards the stream, toppling failures of the stronger bed at the top of the slope and secondary rotational failures of the weaker material in the centre of the landslide mass. There are superb examples of such complex landslide zones on the southern side of the Aguas valley at and east of Los Molinos (*see* chapter 6; Excursion 1), studied by Jim Griffiths and colleagues (*see* above). The weak beds are Tortonian marls; the resistant bed is the sandstone of the overlying Azagador Member, dipping steeply into the valley. The deep incision of the river valley followed the capture of the Feos by the lower Aguas (*see* chapter 4). Downstream of Los Molinos the deep incision has triggered gully and badland development on the opposite side of the valley, in the overlying Abad marl.

Badland erosion is the other major form of modern active slope process (Fig. 5.4; *see* also Fig. 4.4B). The distribution of badlands has already been described (chapter 4). The El Cautivo site in the Tabernas badlands is the most thoroughly researched badland site, with work undertaken by the scientists from the CSIC (Conseco Superior de Investigaciones Cientificas) as well as by Roy Alexander and Adolfo Calvo. The modern badlands there are focused on a series of gully systems incised below pediment surfaces that are cut across the Tortonian marl bedrock. The youngest of these pediment surfaces grades into the Late Pleistocene Tabernas 'lake' (*see* chapter 4: *see* also Excursion 2, chapter 7).

On the gully-side badland slopes there is a strong aspect control of the processes and morphology in the Tabernas badlands, developed on Tortonian marls. The SW-facing slopes are steeper, more or less straight slopes undercut by the gully bottoms (Fig. 5.4B). They are bare of vegetation and have only a very thin weathering mantle over unweathered marl. On these slopes infiltration capacity is low, promoting rapid runoff during rainstorms. This leads to high erosion rates and the development of a dense rill network (Fig. 5.4A). The NE-facing slopes are less steep with an overall concave form. There is a sporadic vegetation cover with occasional bushy plants and annual spring-flowering plants. Importantly, there may be lichens (*see* chapter 7, Excursion 2) on the surface, which both protect the surface and absorb moisture. The surface itself is composed of weathered marl to a depth of up to 10cm above the unweathered marl. During dry weather this weathering mantle is deeply cracked. Infiltration capacity is much higher on these slopes than on the SW-facing slopes,

**5.4** Badland terrain. **A**: Badlands dominated by rill erosion processes, in Tortonian marls, Tabernas basin. **B**: The influence of aspect on badland processes. South and west-facing slopes are steeper, bare and prone to rilling. North and east-facing slopes support some vegetation, are less steep, and erode primarily by small-scale slope failures and mudslides. El Cautivo badland site, Tabernas basin (*see* chapter 7, Excursion 2). **C**: Badlands prone to piping: badlands developed in Gochar-age silts at La Cumbre, southwestern Sorbas basin (*see* chapter 6; other localities).

so that during rainfall much more water is absorbed. This has several effects: first, it promotes at least sporadic vegetation growth; second, it promotes weathering of the marl; third, it promotes shallow mudslides so that erosion is dominated by these processes rather than by overland-flow erosion as on the SW-facing slopes.

One badland process that is not common in the Tabernas badlands, but important elsewhere, is piping (tunnel erosion)(Fig. 5.4C). Piping is common in badlands developed in gypsum-rich marls at several badland sites in the Vera basin. It also occurs in the La Cumbre badland site in the

# Part II

# Keynote sites and itineraries

Suggested selected keynote sites and itineraries, chosen to illustrate the major themes outlined in Chapters 1–5. Cross reference will be made to many of the illustrations in Part I (Chapters 1–5). *Italic script is used for directional information.* Six-figure grid references refer to the Spanish national grid, used on the 1:25000 maps. [For each field site WGS84 (zone 30S) GPS co-ordinates are given in square brackets: Eastings, followed by Northings (note that this format is compatible with Google Earth).]

- ▬ Main faults
- ☐ Quaternary
- ▨ Late Pliocene - Early Pleistocene
- ☐ Late Messinian - Early Pliocene
- ▨ Messinian Gypsum
- ☐ Marls ⎫
- ▨ Reef ⎬ Early Messinian
- ▨ Late Tortonian-early Messinian (Azagador Mbr.)
- ▨ Tortonian
- ▨ Cabo de Gata volcanics
- ▨ Serravallian
- ☐ Alpujarride Schist and Carbonates
- ▨ Nevado - Filabride Schists

**KEY** to all main excursion route and geological maps (Figs. 6.1, 7.1, 8.1, 9.1, 10.1).

# Chapter 6

# The Sorbas basin (Excursion 1)

## Highlights

A transect across the basin, following the Neogene depositional sequence, and at the same time following the Quaternary dissection sequence. (1:25,000 maps: Campohermoso, Polopos, Sorbas, Uleila del Campo.)

The Sorbas basin (Fig. 6.1) serves as a template for understanding the main elements of the regional Neogene stratigraphy and of the Quaternary geomorphology. The basin is bounded by the Sierra de los Filabres to the north and the Sierras Cabrera/Alhamilla to the south. The northern boundary is less tectonically disturbed; there are minor faults but basically it has been tilted. The southern boundary is marked by the strike-slip fault system, which also contains a significant vertical component as a northward thrust. The basin fill contains a complete stratigraphy from the Serravallian to the Plio-Quaternary, initially mostly as marine, then as terrestrial sediments. Sustained uplift during the Quaternary has led to a dominantly erosional landscape, following incision of the major drainage, complicated by a spectacular major river capture, picked out in the river terrace sequence (*see* Fig. 4.5).

The recommended excursion is a south to north transect across the basin, primarily working up the Neogene stratigraphy, and through the dissectional Quaternary landscape. We also recommend a number of other sites within the basin that are not on the general excursion route. The excursion route starts just south of the basin, at Venta del Pobre *(intersection between A9 Motorway and local roads A-101 to Carboneras and AL-104 to Sorbas – 1:25,000 map – Campohermoso: 832946).*

## The excursion route

*From Venta del Pobre* [W2.0674, N36.9928] *to Stop 1: Take the AL-104 north towards Sorbas for about 4km (now on 1:25,000 map – Polopos). The*

**6.4B** View north from Cerro Molatas (Stop 3a) across Los Molinos village to the canyon of the Rio Aguas through the massive gypsum outcrop (centre of photo). Note the landslide on the eastern flank of the valley. Note also the enormous toppled blocks of gypsum in the canyon. Note also the patches of terrace gravels (A,B,C), remnants of previous valley floors of the Rio Aguas. Beyond the gypsum ridge is the Sorbas basin; Sierra de los Filabres form the far skyline.

the greyish tint coincides with the outcrop of the Plio-Pleistocene Gochar conglomerates. In the far distance is the Sierra de los Filabres.

The sequence of Quaternary river incision and river capture can be pieced together from this viewpoint (Fig. 6.4A, *see* also Fig 4.5A). There are terrace gravels capping the gypsum escarpment behind Los Molinos (Terraces A and B), and immediately across the canyon (Terrace B). Terrace C floors the col to your right that marks the abandoned valley of the proto-Aguas/Feos. In places these terraces are marked by well-developed red soils (*see* chapter 4). Piecing together these and other terrace fragments we can reconstruct previous river courses, and estimate the amounts of incision between successive terraces. The total amount of incision at this point since deposition of Terrace A surface (age estimated to be >500ka) is ~160m, of which ~90–100m has occurred since the capture that occurred during Terrace C times (dated at ~70ka). The greatly accelerated incision rate was a direct result of the local base level being lowered by the capture itself. It created an incision wave that propagated upstream through the basin.

One aspect of the rapid lowering of local base level has been to destabilize the slopes above the river. On the outcrop of the highly erodible Abad marl, gullying and badlands developed, for example on the north side of the valley downstream of the capture site (visible to your right).

There are also gullies and badlands in the tributary valley of the Barranco de Los Barrancones that joins the Aguas from the west at Los Molinos. These are better seen from the next stop. In addition, fluvial erosion into weak Tortonian sediments at the base of the slope below you has destabilized that slope. This has created a complex planar landslide (*see* Fig.5.3B), whereby the rigid slab of the Azagador Member has slid downslope over the weak Tortonian marly sediments. In the centre of the landslipped mass secondary rotational failures have taken place, and at the top and sides of the landslip scar toppling failures have occurred in the rigid Azagador sandstones. Similar complex landslides dominate the similar northwest-facing valley-side slope for several kilometres downstream.

If you have time, walk about 400m ENE along the ridge above the landslide scar to where there is a superb view above the abandoned Aguas/Feos valley floored by Terrace C gravels (*see* Fig. 6.4C). The valley is now occupied by the motorway.

**6.4C** View from the north to the capture col, the former valley floor of the Aguas/Feos river. Note the reddish soils on the valley floor and exposed in the far wall of the incision into the valley floor. These soils rest on Terrace C gravels.

*Return to your vehicle and continue north along AL-104, down a series of hairpin bends, through Los Molinos village and up another series of hairpins to where the road levels off on top of the gypsum escarpment. Park on the gravel shoulder: this is Stop 3b (818054)* [W2.0831, N37.0905].

## Stop 3b  Viewpoint above Los Molinos

This spectacular panoramic viewpoint offers superb views of the geology of the central and eastern parts of the Sorbas basin, as well as providing an overview of the site of the late Pleistocene river capture (Figs 6.4C, 6.5A). If you decide to visit only one site within the Sorbas basin, this should be it!

**6.5** Above Los Molinos: the river capture site (Stop 3b). **A**: View south from the (so-called) Eagle's nest (Stop 3b). Note the deeply incised modern valley floor of the Rio Aguas, the landslides and gullied terrain on both sides of the modern incised valley. Note the terrace remnants of the former (pre-capture) Aguas/Feos river, Terrace B forming a spur-top patch to the left of the modern valley, Terrace C forming the abandoned valley floor in the capture col. **B**: Hilltop gravel cap – relatively undisturbed Terrace A gravels, in places incorporating fragmented blocks of Gochar conglomerates at the base, otherwise unconformable on dipping Gochar conglomerates.

Walk ~200m NE up the ridge to your right to the conglomerate-capped hilltop. Beware: there are sheer drops to the right. Take in the panoramic view. Looking southeast you can see the previous site (Cerro Molatas) behind the landslipped slope in the Azagador Member (Fig. 6.5A). To its left is the col marking the abandoned channel of the proto-Aguas/Feos

(Figs 6.4C, 6.5A). Beyond that, and south of the Aguas valley, the slope is formed on the north-dipping Azagador Member with its numerous landslides. Due south from you is the Sierra Cabrera. To the right (west) of Cerro Molatas hill is the gullied terrain cut in Abad marl at the head of the Barancones tributary of the Aguas, beyond which the Azagador scarp crest is capped by gypsum at the top of the tabular hill (El Cerron de Hueli). Beyond is the Sierra de Alhamilla capped by the Cantera reef limestone at Cantona.

Look west along the strike of the gypsum scarp. The gypsum dips steeply north into the Aguas valley. The next hilltop (Los Yesares), ~500m distant, also has a conglomerate cap. To the north, this side of the Aguas valley is a flat spur top, the pre-capture Terrace C. Beyond the Aguas valley is country with a reddish tint, developed on the silts of the Zorreras Member of the Cariatiz Formation. Beyond that is country with a greyish tint, developed on the Plio-Pleistocene Gochar conglomerates. Far beyond is the crest of the Sierra de los Filabres. To the east, across the canyon of the Aguas, is a side view of the gypsum escarpment, capped at its southern end by a patch of Terrace B gravels. Beyond and to the northeast is an extensive gypsum plateau.

Closer at hand, examine the conglomerate cap of the hilltop. We see two separate conglomerates, the lower comprising in part tumbled blocks, in part dipping conglomerate beds, the upper comprising bedded conglomerates deposited around and over the tumbled blocks and dipping conglomerate beds. The tumbled blocks are similar to those on the neighbouring hilltop (Los Yesares). We suspect that these may be Plio-Pleistocene-age fluvial conglomerates of the Gochar Formation, which were either (post-cementation) subject to gypsum solutional (?) collapse from below, or eroded by an incising river. The upper (less disturbed) conglomerate (Fig. 6.5B) we interpret to be a remnant of the first incisional stage, Terrace A. Note the clast content of both conglomerates – abundant green-grey Filabride hornblende schists.

Before leaving this site it is worth returning past your vehicle but crossing the road and walking about 100m towards the SW across a remnant of the gypsum plateau. Here Gypsum karst features, including gypsum tumuli (*see* Fig. 5.2A) are well developed. There is also an impressive view up the valley of Los Barancones to your south of gully systems cut into the Abad marl.

exposed as the cliffs beneath Sorbas village (1km away to your SW). The Sorbas Member is overlain by the uppermost Messinian to basal Pliocene Zorreras Member, the basin-centre facies of the Cariatiz Formation. It comprises red silts deposited in a low-energy coastal plain environment. On three occasions shallow lakes formed, depositing thin white carbonate horizons (*see* Fig.3.8A), though only two are present in the vicinity of Sorbas, visible as white bands within the red Zorreras silts. One of these can be seen from this site, dipping north, within the Zorreras red silts across the valley. The Cariatiz Formation (including the Zorreras Member) culminated in the early Pliocene in a short-lived marine phase (the last occasion when marine conditions were present within the Sorbas basin). The deposits evidencing this marine phase here are bands of yellowish fossiliferous sands, deposited in a nearshore environment (*see* Fig. 3.8B). They are visible from here as a yellowish band above the red Zorreras silts in the valley wall above and to the left of the hamlet of Zoca. (Zoca is the small settlement on the valley floor to your right below this viewpoint.) The Zorreras Member was succeeded by the Plio-Pleistocene Gochar Formation, the grey conglomerates above the Zorreras marine band visible behind Zoca. The whole suite of rocks from the Sorbas Member upwards dips gently to the north.

Cut into the top of the Gochar Formation exposed in the cliff behind Zoca, although not easy to see from this viewpoint, are horizontally bedded gravels of the first incisional terrace of the mid-Pleistocene proto-Aguas/Feos (Terrace A) (Fig. 6.6B; *see* also Fig. 4.5D). The terrace surface (at an elevation of up to 448m) is marked by the almond trees on the skyline across the valley. The flat terrace surface itself is set a little below the lightly dissected end-Gochar surface that forms the terrain to the northwest. The Zorreras viewpoint itself (elevation 449m) is also on Terrace A. There is a horizontal gravel cap over the red Zorreras silts.

Across the valley other remnants of the pre-capture terraces can be identified. Terrace B caps the next hilltop to the south of the almond trees (elevation ~426m), and also caps the isolated mesa on which the town of Sorbas is built (elevation 410m). Within the valley are two gravel-capped hills at ~404m and the spur at the eastern end of Sorbas village at just below 400m, all fragments of Terrace C. The capture at Los Molinos occurred during Terrace C time, and an incisional headcut worked its way rapidly

**6.6B** View west from Zorreras hill (Stop 6.4B). From this viewpoint you can trace the final stages of basin filling, exposed in the valley-side sections, and also the sequence of valley dissection. The view across the valley encompasses the top of the shallow-marine Sorbas Mbr exposed in the base of the valley sides to your left, below Sorbas village. Above this are the coastal plain red silts of the Zorreras Mbr, including the white lacustrine silt bands, exposed in the valley wall opposite you. The Zorreras Mbr culminates in the yellow sandstone marine band, again visible in the cliff opposite, dipping to the north, reaching the valley floor near the hamlet of Zoca to your right. The Zorreras Mbr is overlain by the Plio-Pleistocene Gochar Fm. which forms the valley wall above the Zorreras marine band. This whole suite of rocks dips to the north (to your right). Cut into this sequence are the Pleistocene river terraces that mark the incisional development of the valley. Terrace A cuts unconformably into the Gochar Fm., and forms the flat surface marked by almond trees, across the valley from this viewpoint. The next hilltop to the left comprises Terrace B, cut into the Zorreras Mbr. Two other mesas to the southwest are also capped by Terrace B, including the higher parts of Sorbas village.

The later dissectional stages are also visible from this point, including two abandoned incised meander cutoffs (both of Terrace D age), one immediately below you, the other creating the 'Island in the sky' setting for Sorbas village (see Fig. 6.6C).

upstream, accounting for the deep dissection below the level of Terrace C. Post-capture terraces (we labelled them D1 and D3) are set just above the valley floor, well below Terrace C.

The rapid incision has had another effect on the geomorphology. As well as rapid vertical incision, the high stream power enabled lateral migration to take place. This resulted (here and at many other localities) in the development of incised meandering valley floors. Some of these have cut off, leaving abandoned valley floor meanders, marking the previous river course. There is one of Terrace D1 age just below Zorreras hill, but the most impressive cutoff has resulted in the isolation of the site of Sorbas as an 'island in the sky' (Fig. 6.6C). Originally the western tributary (the Rambla de los Chopos) joined the main headstream of the Aguas (the Rambla de Sorbas) just a little downstream of where the modern bridge crosses the Aguas. However, just north of the site of Sorbas back-to-back meanders, one on each headstream, met and 'captured/diverted' the Chopos into the Rambla de Sorbas, leaving the old valley (that to the south of the town) as an abandoned incised meandering valley. The elevations suggest that this occurred late in the Pleistocene in what we label as D3 time.

**6.6C** The 'island in the sky' situation of Sorbas Village, between the abandoned valley meander of the Rambla de los Chopos to the south and west of the village and the Rambla de Sorbas to the east. Sorbas village itself sits on three terraces, B,C,D (labelled).

*Return to your vehicle.*

If you are interested in soil and calcrete development, there is an excellent roadside exposure of a well-developed red soil and calcrete developed on Terrace B about 300m north of here along the road towards Lubrin (at about 788077). The soil section shows a bright red argillic B horizon above a well-developed pedogenic calcrete. Immediately below the red soil is a laminar calcrete layer below which the calcrete is massive, degenerating with depth into a rubbly calcrete. The parent material (barely visible) consists of gravels of Terrace B.

*If you visit the soil site, either leave your vehicle on Zorreras hill and walk, or if you drive there, you will need to turn afterwards, to head back towards Sorbas. Drive back towards Sorbas, turning right at the main road, but keeping to the main road (through the abandoned incised meander) past Sorbas.*

En route past Sorbas, there are a couple of points to notice. First, on the left-hand side of the sweeping right-hand bend within the abandoned meander, notice the domal structures in the cliff within the Sorbas Member. These are fossil algal stromatolites (*see* Fig. 3.7B: we will be seeing such features at Stop 5). The second point requires a brief stop. Just past the Sol Hotel on the left, take the right-hand turn as though you were going into Sorbas village, but immediately park adjacent to the crash barrier in front of the 'Tanatorio Municipal' building. Walk ~100m north to the overlook above the canyon of the Rambla de los Chopos. From this viewpoint look to the right and you can see how two back-to-back incised meanders, one on the Chopos and one on the Rambla de Sorbas/Rio Alias (Fig. 6.6A) intersected and captured/diverted the Chopos into the Rambla de Sorbas/ Rio Alias, leaving the former valley of the Chopos as the abandoned incised meander. This also had the effect of foreshortening and steepening the long profile of the Chopos, creating the canyon below you.

*Return to your vehicle; rejoin the main road, turning right and heading west out of Sorbas. Continue west for about 1km to a place called Larache. There is a mechanic's workshop on the right. Park on the gravel shoulder just beyond. This is Stop 4c (768064)* [W2.1355, N37.0991].

### Stop 4c West of Sorbas: detailed sedimentology of the Sorbas Member

Walk behind the mechanic's workshop to the edge of the channel of the Rambla de los Chopos. There is a fascinating section exposed in the cliff opposite you (Fig. 6.7A). At the base, within the modern channel, is a clean white, cross-bedded sandstone of the Sorbas Member (Fig. 6.7A; *see* also Fig. 3.7A), which has been interpreted as deposited in a channelled shoreface environment. When you examine the rocks, note the primary current lineation exposed on the sloping bedding plane on the south bank of the modern channel to your right. Note also the beautiful cross-bedding with high angle foresets, showing perhaps a suggestion of current bidirectionality. Might this indicate tidal conditions during the late Messinian? Overlying the cross-bedded sands are laminated clays, in which bird footprints have been found (lagoonal conditions), in turn overlain by low-angle cross-bedded sands (aeolian dunes). These are overlain unconformably by Gochar conglomerates. Cut into these conglomerates is a late Pleistocene river terrace gravel (Terrace C), within which are blocks of the cemented Gochar conglomerate (Fig. 6.7A). It is interesting to note that the Zorreras Member is missing from this sequence. This may relate to its erosional removal before deposition of the Gochar conglomerates, or perhaps to non-deposition on what, at the time, was a topographic high.

*Return to your vehicle. Rejoin the main road, heading west, but just after the bridge over the Rambla turn right onto the AL-813 road towards Uleila. Follow this road north for about 3.2km. There turn right onto a minor road,*

**6.7** West of Sorbas. **A**: Section at Lararche (Stop 6.4C), west of Sorbas. The floor of the Rambla de los Chopos cuts into beautiful cross-bedded sands of the Sorbas Mbr, possibly showing some suggestion of bidirectionality (maybe tidal?). At the downstream end of the section are well-preserved primary current lineations. The cross-bedded sands are overlain by lower energy silts, in which fossil bird footprints have been found. These are, in turn, overlain by low-angle cross-bedded sands, suggesting dune deposition. The whole sequence indicates a final Sorbas-age marine regression, shallow-marine sands, overlain by lagoonal muds then by aeolian dunes. The upper part of the section is capped unconformably by terrace C gravels. Note the blocks of cemented Gochar conglomerates at the base of the terrace gravels. These have been derived from collapse of the nearby valley wall during incision. **B**: Section cut by the Rambla de Sorbas near Gochar village showing Pleistocene Terrace B gravels resting unconformably on dipping Plio-Pleistocene Gochar Formation conglomerates.

*heading east towards Gochar. After about 1.7km turn left into the valley, across the Rambla and up through Gochar village. Continue, now heading north for another 2km towards Moras, finally turning right at a T-junction for the last 1.2km into Moras Village. Park at the entrance to the village. This is Stop 5a. (774109),* [W2.1311, N37.1399].

En route from Larache· crossing the two ramblas note the section in dipping Gochar red silts and conglomerates to your right. You then climb up onto the little dissected end-Gochar depositional surface until after you make the right turn towards Gochar village. In this area you then drop onto the Aguas terraces above the Rambla de Sorbas. Terrace B is particularly well developed here. When crossing the rambla, where the Rambla de Moras and the Rambla de Gochar meet to form the Rambla de Sorbas, note the section on the right (Fig. 6.7B). Horizontal terrace gravels rest unconformably on dipping Gochar conglomerates. Note also the faulting of the Gochar Formation here. There is also a large section, cut by the incising Rambla de Gochar south of Gochar village, exposing Gochar conglomerates. Beyond Gochar village you are back onto the end-Gochar surface along the spur top before you drop down into the valley of the Rambla de Moras at Moras village.

## The northern margin of the basin

### Stop 5a  Moras

Moras is a delightful little village, set above the incised meanders of the Rambla de Moras, near the northern margin of the basin. The geology is superb. Exposed by the incision, are rocks and structures ranging from the Messinian Cantera Member reefs to the Plio-Pleistocene Gochar conglomerates (Fig. 6.8A,B). Likewise, the Quaternary and the geomorphology are fascinating, exhibiting interesting contrasts in morphology during incision.

Walk 200m north along the goat track that skirts an arcuate amphitheatre. Take in the view from here. You are on rocks at the top of the Sorbas Member, here exhibiting algal stromatolite domes (*see* Fig. 3.7B) within silty limestone beds and conglomerates. Looking back towards the village, note that there is an intraformational unconformity within the Sorbas sequence (Fig. 6.8A). This indicates a growth fold, active during sedimentation of the Sorbas beds. Further evidence of ongoing deformation on the growth fold can be seen in the cliff opposite the village (Fig. 6.8B). That cliff exposes

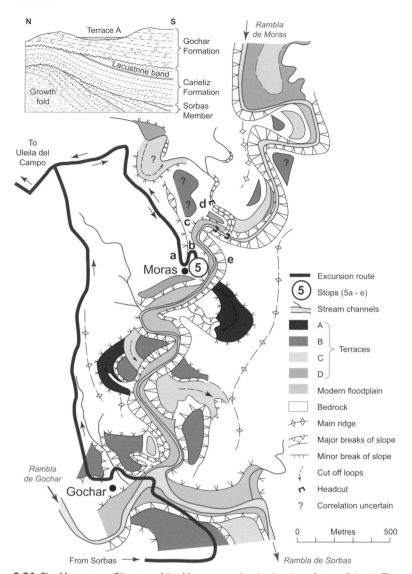

**6.8A** The Moras area. Site map of the Moras area, showing location of stops 5 (a–e). The inset shows the geological structure. Note the successive intra-formational (within the Sorbas Mbr) and inter-formational unconformities, indicating the progressive development of a growth fold with an E–W axis through Moras.

the Cariatiz Formation (Moras Member, the basin-marginal equivalent to the Zorreras Member, seen in the basin centre), within which one of the white lacustrine marker bands is visible. Within the Moras Member there is an angular unconformity above which the top of the Moras Member is

**6.8B** The cliff behind Moras village (to the east), composed of dipping Sorbas Mbr sandstones at the base, overlain in turn by Cariatiz Fm. (Moras Mbr) conglomerates, including at least one of the white silt lacustrine bands. The Moras Mbr culminates in a beach conglomerate and is overlain by Plio-Pleistocene Gochar conglomerates. To the right is the incision into the Gochar of a meander of Terrace A.

marked by the yellowish marine band, here conglomeratic, rather than the sandy facies evident further south. Overlying the Moras Member are the coarse conglomerates of the Gochar Formation, showing an overall dip to the SE. Right at the top of the cliff are gravels of Terrace A, set into the Gochar conglomerates, capped by a deep red soil (*see* below for a discussion of the geomorphology).

Continue along the goat track, which crosses a low watershed into a small tributary valley, then scramble down the stream bed. The stream bed crosses Messinian reef limestones (Cantera Member) which are unconformably overlain, over an eroded surface, by Sorbas-age conglomerates and stromatolite domes (*see* Fig. 3.7B). These are in turn overlain by reddish sands and conglomerates of the Moras Member. (Note the presence of the lowest of the white carbonate lacustrine marker bands). Where the tributary barranco meets the Rambla de Moras, note the beach conglomerate, with attached fossil barnacles. The conglomerate forms the dry waterfall in the main channel. This is the top-Moras marine band. Above it and outcropping in the cliff to the south of the main rambla, are Plio-Pleistocene Gochar conglomerates.

Make your way back to Moras village and your vehicle – best to scramble up the 'amphitheatre' slopes. Before leaving Moras, note two points.

First, above the village and exposed in the roadcut on the last hairpin bend is a channel cut into Moras-age conglomerates. It is filled with lagoonal carbonate muds, with a warm-water brackish micro-fossil assemblage, indicating a coastal situation: another expression of the early Pliocene marine incursion (the last one) into the basin.

Second, a note on the geomorphology. As with most valleys within the Sorbas basin, the Moras valley is characterized by incised valley meanders. There is, however, an interesting contrast between the valley upstream and downstream of Moras village (Fig. 6.8A: map). By piecing together the various terrace fragments, previous courses of the valley can be reconstructed. Upstream of Moras incision has been fairly simple. The terraces offlap one another (Fig. 6.8C); successive incisions have undercut the outsides of the valley bends, resulting in ingrown meanders and a progressive increase in valley sinuosity with time (Fig. 6.8D). In contrast, downstream of Moras previous courses have been much more tortuous, involving numerous cutoffs, leaving perched abandoned meanders. Lateral activity during incision was obviously greater here than it was upstream. This is presumably a reflection of higher stream power. This, in turn, probably relates, at least in part, to steeper stream gradients downstream of Moras. The incision wave that followed the Aguas/Feos capture peters out upstream of Moras, so this may partly explain the steeper gradients

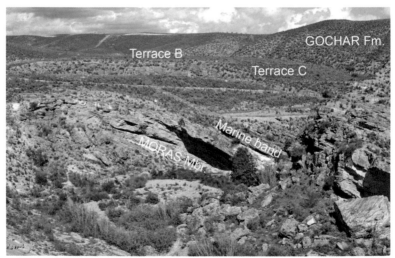

**6.8C** Photo looking upstream from Moras village. The incised meanders have become progressively ingrown during incision.

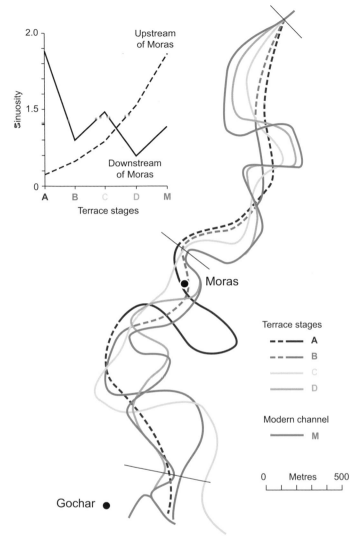

**6.8D** Reconstructed incision sequence from terrace A to the modern valley. Note how, upstream of Moras (i.e., upstream of the growth fold) the incised meanders have become ingrown: sinuosity progressively increases, whereas downstream of Moras numerous cutoffs have occurred.

downstream. However, the contrasts relate particularly to Terraces A and B, which clearly pre-date the capture. Most likely there is a tectonic control. Downstream of Moras the valley is aligned down the flank of the growth fold. The growth fold was active well after deposition of the Gochar Formation; activity continued well into the Pleistocene. The effect of that activity would be to steepen stream gradients downstream during incision, accentuating erosional activity. In contrast, upstream of Moras the effects would be to reduce erosional activity.

*Leave Moras heading west (there is no other option!), but at the first road junction DO NOT turn back south towards Gochar, but continue west to join the Sorbas–Uleila road (AL-813). At that junction turn right to head northwest for about 6km to Uleila, to complete your transect of the Sorbas basin. With Uleila village to your left, just short of the village turn right onto the C-3325, a road that heads north to cross the Sierra de los Filabres. Continue along this road for about 800m to the outside of a left-hand bend around a spur. Park on the gravel shoulder on the right. This is your final stop 5b (712167)* [W2.1998, N37.1926].

En route from Moras, note the large section in Gochar conglomerates (the Gochar type-section: *see* Fig. 3.10) on the left just at the junction of the Moras road with the AL-813, Uleila road. Along the Uleila road you leave the outcrop of the Gochar Formation, and cross that of the Cantera reefs onto the basement in the mountain-front zone near Uleila.

**Stop 5b  Above Uleila**

You are now on the mountain front of the Sierra de los Filabres. Note the basement rocks exposed in the roadcuts are medium grade metamorphics, the grey hornblende schist that dominates the clast composition of the fluvial deposits of the Sorbas basin, from the Gochar conglomerates to the modern stream gravels. Note that the mountain front of the Filabres is an erosional mountain front, very irregular in outline, in complete contrast to the linear fault-bound northern mountain front of the Sierra Cabrera/Alhamilla seen earlier. Note also how the mountain front is characterized by fairly small modern alluvial fans (Fig. 6.9), backfilling into the mountain catchments behind. Below you to the east is one such feature – planted with almond trees.

***End of Excursion 1***
*******************************

# Other localities in the Sorbas basin that merit a visit

There are several other localities in the Sorbas basin that are well worth a visit, depending on your interests. These are (i) the Cariatiz area, especially

**6.9** Uleila del Campo: the passive mountain front of the Sierra de los Filabres. Note how the mountain front is irregular, with alluvial fans backfilling into the valleys.

the Messinian (Cantera) reef; (ii) a hike from Hueli to the summit of Cantona, for panoramic views; (iii) Urra, the site of local late Quaternary deformation; (iv) the gypsum plateau in the east of the basin (gypsum surface features); (v) the Lucainena valley in the west of the basin and a drive through the Sierra Alhamilla; (vi) La Cumbre badlands in the south of the basin.

### Cariatiz – especially the geometry of the Messinian (Cantera) reefs
*It is best to approach Cariatiz from the west. Take the Sorbas to Lubrin road north for about 7.5km to an intersection on the right (at 788127) to Cariatiz. Turn right and drive for about 1km to a viewpoint above a deeply incised valley on your left/east (795116)* [W2.1052, N37.1446].

En route from Sorbas, after passing Zorreras hill (*see* Excursion 1, Stop 4b) the road runs up the end-Gochar surface, then across that surface to the viewpoint. At the viewpoint you are essentially near the apex of a large Gochar-age alluvial fan complex, fed from the Sierra de los Filabres from the north. In front of you the Barranco de los Castanos has cut through the margins of the fan, diverting the drainage to the east (*see* map, Fig. 6.10A). Across this valley is a spectacular view of the Messinian (Cantera Fm.) reef *in situ* (Site 1 on Fig. 6.10A: see also Fig 6.10B, compare with Fig. 3.5B). It is built on the Filabres basement, with the prograding reef platform that

91

*tank (796026). Park here* [W2.1066, N37.0652] – *the track is impassable beyond this point.*

Walk from here, following the main track which turns south-west, below the abandoned village of Hueli. You are now stratigraphically below the gypsum, which forms the scarp-top hill of El Cerron de Hueli a couple of kilometres away to your left (east). You are stratigraphically within the Abad marl, the basin-centre lateral equivalent of the Cantera reefs. As you walk SW from Hueli something of this lateral transition becomes evident. The low hills to your left are patch reefs within the Abad marl. Soon, as you approach the steeper part of the slope, the patch reefs coalesce to form a continuous reef limestone horizon, which itself overlies the Azagador calcarenite.

At this point (788019) it is worth taking a slight diversion. The track veers to the right up a spur. We suggest you keep to the bottom of a small valley. After ~300m you will emerge onto a col area above the Mizala valley. The Mizala is a subsequent stream cutting back from the Feos at Penas Negras (*see* Excursion 1, Stop 2) along the strike of the Tortonian marls and turbidites, that are capped by the scarp-forming Azagador Member along the northern valley side. The southern valley side is marked by the northern Alhamilla boundary faults. At this point, deep gullies, cut into the Tortonian rocks, preferentially along the steeply dipping marly horizons, are extending the Mizala drainage towards the west. These gullies are about to intercept the line of drainage you have been following upstream, and to bring about a capture of Sorbas basinal drainage by Mizala/Feos drainage (*see* Fig. 6.3).

Return to the track and continue upwards. Note that the track is now more or less on the Tortonian/Azagador unconformity. Incipient gullying on the track is preferentially attacking the weaker Tortonian marly horizons. At the col overlooking the Lucainena valley to the west (782014), leave the track and climb upwards to the left to the summit of Cantona.

The panoramic view from here (Fig. 6.12 A,B) includes the skyline of the Sierra de los Filabres to the north. Far to the west is the Sierra Nevada. In front of the Filabres are (west to east) the Tabernas, Sorbas and Vera basins. Looking east, the Sierra Cabrera can be seen beyond the transverse course of the Feos valley (the course taken by the proto-Aguas/Feos). Nearer is the summit of the eastern part of the Sierra de Alhamilla, bounded to its north by a major fault. We are standing on rocks of the Cantera (reef)

**6.12** The panorama from Cantona. **A**: View west into the Lucainena valley. The Lucainena fault marks the thrust forward front of the Sierra de Alhamilla. The valley is drained by the Rambla de Lucainena, the main headwater of the Rio Alias. It is a transverse drainage crossing the Sierra de Alhamilla into the Almeria basin to the south. **B**: View south, across the Sierra de Alhamilla to the Almeria basin (marked by the 'plastic agriculture'), bounded by La Serrata, a ridge of volcanic rocks along the Carboneras fault system. In the far distance is the main mass of the Cabo de Gata volcanics.

Member that have been arched upwards as a result of the uplift of the Sierra de Alhamilla. To the west is the main mass of the Sierra de Alhamilla; beyond is another transverse drainage, the Rambla de Lucainena, the main headwater of the Rio Alias, that drains into the northern portion of the Almeria basin. The Almeria basin lies to the south and east of the Sierra de Alhamilla, distinct in its cover of 'plastic agriculture'. To the south, beyond the Almeria basin, lies the Carboneras fault system, marked by a line of low hills (La Serrata) formed on Cabo de Gata volcanics, beyond which lie the low peaks of the main mass of the volcanic Sierra de Cabo de Gata.

### Urra: site of local Late Quaternary deformation

*Urra is a privately run field centre. Before accessing this site you should seek permission from the owners. Urra is on the AL-104 Sorbas to Penas Negras road about 2.5km east of Sorbas village. The track to the field centre is on the left of the road from Sorbas, just over a kilometre beyond the bridge over the Aguas. Park at the field centre (803062)* [W2.0965, N37.0958].

At Urra the post-capture terraces of the Rio Aguas are exceptionally well developed (Fig. 6.13A) We have labelled them (oldest to youngest) D1, D2, D3. Terraces D1 and D3 are recognizable throughout the modern river systems, but Terrace D2 is local to the Urra area. Terrace D2 sediments show intense deformation.

The setting of Urra is important in an evaluation of the possible causative mechanisms for the Terrace D2 deformation. The area is bounded to the south by steep, northerly-dipping gypsum. The area itself is underlain by marly rocks of the overlying Sorbas Member, but gypsum knobs, which appear to be *in situ* (?), are exposed on the bed of the Rio Aguas to the northeast of Urra. This suggests an E–W syncline underlying Urra.

Urra also lies only about 4km upstream of the late Pleistocene Aguas/Feos capture site. It is where the Barranco de Hueli tributary meets the main Rio Aguas. The post-capture incision wave on the Aguas must have removed an enormous amount of relatively easily erodible material. The immediately pre-capture (locally faulted!) Terrace C about 1km to the east of Urra (Fig. 6.13B) now reliably dated to ~70ka) is at an elevation of ~380m, whereas the elevation of the modern Aguas at Urra is only ~310m. Set below Terrace C, NE of Urra, is Terrace D1 (Fig. 6.13C: elevation ~350m), which probably dates from about 40ka. It comprises several metres of gravel and silts, cut into rocks of the Sorbas Member and culminating in a moderately developed red soil (*see* Fig. 4.7D). Following Terrace D1 there was a major incision to modern river level and below, then aggradation of gravels and silts in the Aguas, but silts in the tributary barranco almost to the level of Terrace D1, to form Terrace D2. There is only very weak soil development on the D2 surface. Furthermore, D2 sediments are intensively deformed (Fig. 6.13D). Another incision followed to a few metres above the modern river level, followed by the aggradation of Terrace D3 (undeformed, and estimated to date from the latest Pleistocene 15–10ka, and with almost no observable soil development on its surface). Since then, Terrace D3 has been dissected and the lowest (Holocene) terrace and modern floodplain have been formed. The post-Terrace D2 incision produced a highly tortuous tributary

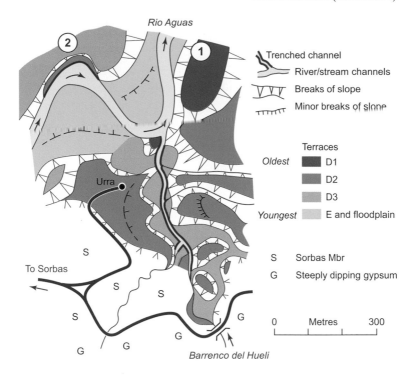

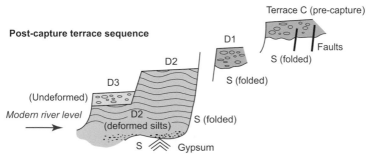

**6.13A** The deformation at Urra. The Urra area, showing the D1,2,3 terrace sequence on the Barranco de Hueli. Also shown are the locations of the suggested field sites 1,2. The summary sequence is shown below.

barranco, since when there have been cutoffs resulting in abandoned valley meander loops (Fig. 6.13A,B). The channel straightening may have been anthropogenically induced during the Holocene.

To view this sequence, walk first to the promontory east of the field centre (Fig. 6.13A). From there, looking east, you can see the Terrace D1 remnant capping the hill in front of you and the great thickness of D2 silts

**6.13B** General view of the Urra area, looking east. The right skyline is formed by northerly dipping gypsum, the left skyline by Terrace C of the Rio de Aguas. In the middle ground is an abandoned meandering valley of the Urra drainage, exposing the thick silts of Terrace D2.

**6.13C** View east from Cortijo Urra. In the middle distance are the buff silts and gravels of Terrace D1, capping slightly deformed Sorbas Mbr. siltstones. In the foreground are buff silts of Terrace D2 cut deeply into Sorbas Mbr. siltstones

to your front right. The undeformed D3 terrace forms the dissected valley floor of the barranco and the well-defined low terrace surface across the Rio Aguas. From here, walk down into the Aguas channel and head for the exposure of the D3 terrace you could previously see. Closer to, you will see that the terrace surface is on undeformed gravels of the Rio Aguas,

**6.13D**  Section cut by the Rio Aguas. At the base are deformed D2 sediments of the Rio Aguas, capped by undeformed D3 sediments.

below which are highly deformed gravels and silts. The undeformed sediments are the D3 terrace; the deformed sediments below are the deformed Terrace D2 sediments of the Aguas.

Return to the field centre, then if possible (in recent years this area has become very overgrown!) follow the barranco upstream. There are several small streamside sections showing deformed D2 sediments over Sorbas marls. About 200m upstream there is a large section on your left (east), showing a 10m stack of slightly deformed laminated muds overlain by barranco gravels. Further on, near the road bridge, there is another large section showing a mixture of wet debris-flow deposits and bedded silts that show synsedimentary thickening towards the barranco channel, again more evidence of synsedimentary deformation (Fig. 6.13D).

The large amount of pre-Terrace D2 incision is easy to explain, in relation to an incision wave propagating upstream from the Aguas/Feos capture site, but the massive aggradation of the D2 terrace, together with the contemporaneous, but time-limited, deformation are less easy to explain. *See* Fig. 6.13A for a summary of the sequence. We see three possible explanations: (i) the rapid incision and associated offloading may have caused instability in the underlying gypsum, and diapiric deformation (gypsum is a relatively light mineral and could respond to offloading by rising); (ii) gypsum is a soluble mineral and its dissolution could have caused karstic collapse of the overlying material; (iii) Urra is on the line of an important tectonic lineament in the underlying basement and tectonic activity could have been involved. None of these three potential explanations is entirely satisfactory. Maybe a combination of several possible causes is the answer.

### The Lucainena valley and the transverse course of the headwaters of the Rio Alias

There is iron ore within the uppermost Nevado–Filabride (?) basement complex at Lucainena. There were mines up the slope from the northern boundary fault of the Sierra de Alhamilla (for a general view of the Lucainena valley *see* Fig. 6.12A). At the peak of production about a century ago, a railway was built from Lucainena to the port of Carboneras on the coast. The railway has long since been abandoned, but the (now surfaced) trackbed provides a fascinating excursion drive through the canyon of the Rambla de Lucainena (*see* Fig. 5.5A), the main headwater of the Rio Alias (*see* chapter 9, Excursion 4), through the Sierra de Alhamilla. The Rambla de Lucainena, a transverse drainage, whose origins were presumably similar to those of the proto-Aguas/Feos (initially superimposition, followed by some degree of antecedence: *see* chapter 4), has extended its own headwaters westwards as a subsequent stream across the southern margins of the Sorbas basin. This is similar to the development of the Rambla Mizala (*see* Cantona hike, above).

*To access the start of this route, drive to Lucainena village, turning south off the Sorbas to Tabernas main road about 7km west of Sorbas (713044) onto the AL-130.*

This road to Lucainena crosses a col in the north-dipping Azagador escarpment onto the Tortonian marls and Turbidites of the Lucainena valley. Lucainena village is on the south side of the valley, right up against the steep mountain front of the Sierra de Alhamilla, coincident with the northern boundary fault of the Sierra de Alhamilla (Fig. 6.12A). Continue past Lucainena. The road skirts the south side of the valley and begins to climb onto the Sierra de Alhamilla. About 3.5km beyond Lucainena, at Rambla Honda (746990) take a left turn onto a road/track that follows the Rambla Honda down to El Saltador, where you enter the canyon of the Rambla de

Lucainena. Note the massive stream-bed calcrete forming a (dry) waterfall here. From here on you will follow the trackbed of the old railway. The track is initially on the south side of the canyon but later crosses to the north side. The canyon walls (*see* Fig. 5.5A) are initially in basement schists, but later these are unconformably overlain by the the Azagador Member and Cantera reef limestones. Within the canyon are patches of river gravels, but we do not know the details of the terrace sequence here. At one point you drive through a short tunnel, then drop down onto the rambla floor. From there the best-maintained track climbs up the north side of the valley to Polopos. We will describe the geology and geomorphology of the Polopos area in chapter 9 (Excursion 4). There is a road out from Polopos to the AL-104, north of Venta del Pobre (*see* Excursion 1).

### The gypsum plateau east of Sorbas, above the Aguas valley at La Herreria

The primary purpose of this side trip is to examine the detailed features related to processes on gypsum surfaces. Be warned and be careful. The area is a protected area and the features are easily damaged. There is driving access along the track to La Herreria, but you should not walk off the track.

*Access: On the Sorbas to Vera main road, about 8.5km east of Sorbas (832104), turn right onto a track that runs SSE across the gypsum plateau, past the gypsum quarry, for about 2.5km where you are on open gypsum plateau surfaces (854082),* [W2.0382, N37.1110].

You are now in a zone where the weathering features on the gypsum are particularly well developed (*see* Fig. 5.2A,B). The gypsum formation includes both the anhydrous and hydrated forms (anhydrite and gypsum respectively). During rainfall, the anhydrite absorbs moisture and hydrates to form gypsum. Gypsum has a greater volume, but lower density than anhydrite (*see* chapter 5), so on absorption of moisture it expands. This creates surface 'blisters' (or so-called tumuli – *see* chapter 5, Fig. 5.2A) which range in scale from a few centimetres in diameter and height to much larger forms maybe 50cm high and more than 1m in diameter.

While at this location it is worth driving another kilometre or so to the lip of the Aguas valley above La Herreria (854076). The view is spectacular, and relates to the rapid incision that led to the Aguas–Feos capture (see chapter 4; chapter 6, Excursion 1). The gypsum forms a caprock on this side of the valley. Below you, on this side of the valley, are extensive badlands cut into the underlying Abad marl. On the far side of the valley is the landslipped terrain where the steeply northward-dipping Azagador sandstones rest unconformably on weak Tortonian marls and turbidites. The whole hillside is prone to complex landsliding. Perhaps it is significant that, after the initial possibility of locating the motorway on the south side of the valley, the decision was eventually to locate it on the north side.

## La Cumbre

*Access: From Sorbas drive ~1km west towards Tabernas. At Lararche (the site of Stop 4c on the main excursion), turn left (south) onto a track between the houses. Bear left by a quarry entrance, onto an arcuate plateau cut into the Sorbas Member. This is probably a former meandering valley of the Aguas/Feos (Stage A). From here the track bears right obliquely up the hillslope towards La Cumbre. Pause at top of the rise for the view north over Sorbas and the centre of the Sorbas basin towards the Sierra de los Filabres. The track continues along a watershed, crossing a series of cols that represent former SW–NE drainage lines that were captured by the steep aggressive S–N drainage to your right (west), the Barranco de Mocatan and its tributaries. The captures took place in response to the deep incision of the Aguas in the Sorbas area following the Aguas–Feos capture (see above). The track then emerges above a deeply dissected area to the right (west). These are the La Cumbre badlands (767042)* [W2.1364, N37.0796].

### Badlands cut in Plio-Pleistocene silts

La Cumbre badlands (*see* Fig. 5.4C) are cut in Plio-Pleistocene Gochar Formation distal sands and silts, derived from the southern margin of the basin (*see* below). The badlands themselves are developed on the side slopes of west-flowing tributary valleys of the north-flowing Barranco de Mocatan stream system. They show a marked aspect-related asymmetry, with well-vegetated north-facing slopes but bare, erosional south-facing slopes. As with the El Cautivo badlands in the Tabernas basin (*see* chapter 7, Excursion 2, Stop 5b), some of the asymmetry may be due to a structural control, but nevertheless is expressed in the erosional processes on the badland slopes. One characteristic of the La Cumbre badlands is the importance of subsurface piping processes (*see* Fig. 5.4C), which may be due to the presence of soluble salts within the sands and silts. Piping is evident on the badland slopes themselves, but also in the valley fill below the badlands.

## Plio-Pleistocene (Gochar Fm.) sediments and palaeogeography on the southern margin of the Sorbas basin

The Gochar Formation basin-fill sediments on this southern margin of the basin are very different from those in the basin centre and on the northern margins. Several reasons can be put forward to explain these differences. The source area for these sediments was on the margins of the Sierra Alhamilla and over intermediate ground dominated by Tortonian turbidites and sandstones The original drainage here was from south to north, supplying conglomerates fed by Triassic metacarbonate rocks of the Sierra de Alhamilla. Later development of the strike-orientated Lucainena drainage cut off this source area so that the bulk of the sediment sequence is dominated by sands fed from the local Tortonian sandstones and Messinian reef carbonate clasts. The great thicknesses of the Gochar-age sandy sediments in this area is due to the position here of a major active tectonic lineament (the same lineament that passes under Urra, see p. 100) that caused diversion of sediments into topographic lows, local unconformities and overturning of sediments, all locally accentuating sediment thickness.

# Chapter 7

# The Tabernas basin (Excursion 2)

## Highlights

The Tabernas basin is the basin with the maximum post-Pliocene uplift. However, through the Rioja corridor it is linked to the western end of the Almeria basin with the least uplift. Exposed within the Tabernas basin is the most complete Serravallian and Tortonian sequence. Little remains of any Messinian sediment, but there are thick Pliocene fluvial and fan-delta conglomerates in the west, linking through the Rioja corridor to the Almeria basin. Furthermore, the steep, tectonically induced gradients, together with ongoing tectonic activity during the Pleistocene, have combined to produce the most remarkable contrasts between the intensely erosional landscapes in the western part of the basin (the badlands area) and the depositional landscapes of the eastern part of the basin (coalescent alluvial fans). The basin is bounded to the north by the uplifting, but now tectonically rather passive, Sierra de los Filabres, and to the south by the active faulted mountain front of the Sierra de Alhamilla. To the east the basin links to the Sorbas basin.

The suggested excursion starts in Tabernas village (Fig. 7.1) with a spectacular overview of both eastern and western parts of the basin (Fig. 7.2), followed by field visits in turn to the east and west parts of the basin. (1:25000 Maps: Tabernas, Lucainena de las Torres, Arroyo de Verdelecho, Los Yesos, Gador.)

## Tabernas basin – central area and panorama from Tabernas village

*The excursion starts at Tabernas castle. From the Sorbas–Almeria road (the old N340), take the eastern exit for Tabernas village. About 150m beyond the junction take the minor road to the SW and bear right towards the Sports centre and park [W2.3926, N37.0532]. From the car park, follow the path up to the*

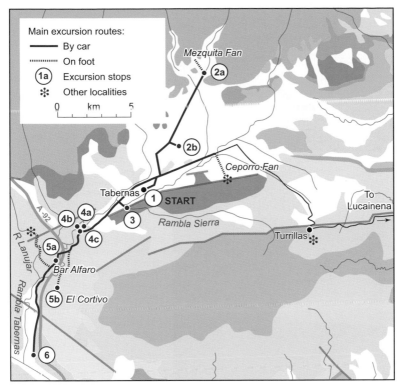

**7.1** Tabernas basin – geological map, excursion route and stop locations.

*castle ramparts. Once, there was an Arabic castle here, but the modern castle is made of breeze blocks – it was made for the Tabernas 'spaghetti western' film industry – for the film El Cid. This is stop 1 (539011)* [W2.3951, N37.0522].

## Stop 1 Tabernas castle: overview

Despite its lack of authenticity, Tabernas Castle offers a superb panoramic view over the Tabernas basin (Fig. 7.2A,B,C). Looking east (Fig. 7.2A) the basement mountains of the Sierra de los Filabres are on the left and those of the Sierra de Alhamilla are on the right. In front of the Sierra de Alhamilla is a ridge that extends westwards towards the south of Tabernas village. This is the Serrata del Marchante, an anticline in Serravallian conglomerates that has been uplifted and thrust forward by a major fault parallel with the northern boundary fault of the Sierra de Alhamilla. Again, looking east, the low terrain between the Sierra de los Filabres and the Serrata del Marchante comprises coalescent Quaternary alluvial fans

**7.2** Stop 7.1: views from Tabernas castle. **A**: View to the east, the aggrading part of the basin. The smooth terrain around the tower of the solar power plant comprises the coalescent Quaternary alluvial fans fed by the Sierra de los Filabres, the mountains in the background. The roadcuts in the foreground expose the underlying Tortonian rocks, overlain locally by the Tabernas lake sediments. **B**: View to the south across Tabernas village and the deeply entrenched Tabernas canyon to the Serrata del Marchante (a thrust forward anticline in Serravallian conglomerates). The skyline is the Sierra de Alhamilla. **C**: View to the north-west, including calcreted, gravel-capped mesas forming a high terrace above Tortonian rocks. Below them are the Tabernas lake sediments capping Tortonian rocks. To the left (SW, not shown in the photo) is the deeply dissected badland part of the basin the 'desierto de Tabernas'.

emanating from the two ranges. Those on the left, around and behind the tower of the solar power station, are large low-angle fans sourced by large catchments within the Filabres. Those to the right are smaller, steeper fans derived from small catchments within the Serrata del Marchante. Between the two is a large low-angle fan derived from the Sierra de Alhamilla, via the Rambla de Nurias, a stream that wraps around the Serrata del Marchante. The skyline to the east behind the Filabres fans is formed by the escarpment of the Azagador Member capped by local gypsum, in the area that marks the divide between the Tabernas and Sorbas basins. The level low ground in the foreground marks the upstream limit of the sediments of the late Pleistocene Tabernas 'lake' (Fig. 7.3: *see* also chapter 4). The 'lake' is actually in two parts, that upstream of Tabernas canyon (visible from this

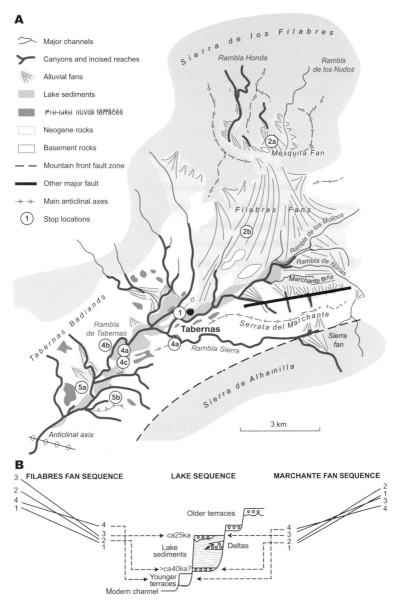

**7.3 A**: Map of the Tabernas fans and 'lake' system, showing stop locations. **B**: Schematic cross-section illustrating the relationships between terraces, fan phases and the 'lake' sediment sequence.

viewpoint to the south of Tabernas village: Fig. 7.2B) forming a surface at elevations of ~430–440m., whereas that downstream of the canyon forms a surface at elevations 320–330m.

Looking south, across the rooftops of Tabernas village (Fig.7.2B), you can see the deep canyon of the Rambla de Tabernas, finally incised after drainage of the Tabernas 'lake'. Beyond is the ridge of the Serrata del Marchante, behind which is the crest of the Sierra de Alhamilla.

To the west is the 'Desierto de Tabernas' (Fig. 7.2C; *see* also Fig. 4.4B), the badland terrain developed on thick Tortonian marls and turbidites, bounded to the right (northwest) by the Sierra de los Filabres. In the far distance to the northwest are the Sierra Nevada. The badlands are bounded to the south by the faulted front of the Sierra de Alhamilla. The higher hills in the centre of the badland area are mostly formed of dipping sandstone beds within the Tortonian turbidites. The flat surfaces in the foreground and in the centre of the middle distance mark the late Pleistocene Tabernas 'lake' sediments (Fig. 7.3). These are trenched by the modern river channels.

*Return to your vehicle. Drive east out of Tabernas village. At the edge of the village (546016) turn left onto the A-349, towards Senes and Tahal. Cross the main road and continue north for about 8km, keeping on the A-349 towards and past the solar power plant. You are now approaching the very irregular Filabres mountain front. The road crosses a large channel on the surface of the Rambla Honda fan and passes a low hill of black rock on the left. About 200m beyond that hill, park on the section of old road on the left (582088). This is stop 2(a): Mezquita fan* [W2.3472, N37.1220].

## Tabernas basin – eastern part: Filabres mountain front and coalescent Quaternary alluvial fans

### Stop 2a  Mezquita fan

En route from Tabernas you have been driving over the extensive surface of a large late-Pleistocene alluvial fan complex, fed from the Filabres by the valleys of the Ramblas Honda and los Nudos. You are now in the Nudos valley at the foot of a small tributary alluvial fan, the Mezquita fan (Fig. 7.4A,B,C). The Filabres mountain front is highly irregular, tectonically passive, uplifted and tilted rather than faulted. It is defined by burial by backfilled large and small alluvial fans. Walk NNW up the distal surface of the alluvial fan and into the fanhead channel at the intersection point, where the modern channel issuing from the fanhead trench meets the fan surface. Follow the fanhead-trenched channel almost to the mountain front, ignoring the first, rather degraded, section, then curving to the right, to where there is an excellent section in the fan deposits (579107).

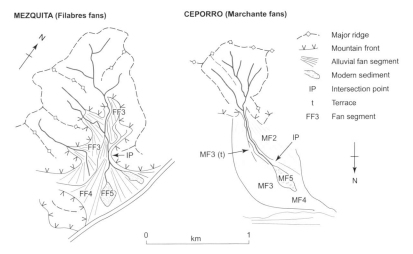

**7.4A** Mezquita fan and Ceporro fans: maps (S).

**7.4B** Mezquita fan and Ceporro fans profiles.

**7.4C** Mezquita fan – general view upfan from near the intersection point. Note: on the right of the channel in the middle distance the older fan surface.

Before examining the fan deposits, examine the clasts within the deposits. They are graphite mica schists of the Tahal unit of the Filabres complex (*see* chapter 2; *see* also Fig. 2.1A). They are highly fissile, rich in graphite, but also contain pyrite.

to pre-date or correlate with the early phases of the formation of Tabernas 'lake'. Unit F3 appears to relate to the final phases of the lake (culminating at ~20ka; (Fig. 7.3, *see* also chapter 4). The development of the modern fanhead trenches and deposition on the distal fan surfaces would therefore appear to relate to the last 20,000 years.

*Return to Tabernas. At the main road, turn right, bypassing Tabernas village. About 2.5km from that junction you cross the bridge over the Rambla de Tabernas. Immediately over the bridge, turn left onto a track down to the floor of the rambla. Turn right on the Rambla bed and drive upstream for about 300m to the junction with the Rambla Sierra, coming in from the right (south). Turn into the Rambla Sierra and park under the shade of trees on the rambla floor. Rambla Sierra is Stop 3 (S34007), [W2.3997, N37.0413].*

## Tabernas basin: western part – badlands cut in Tortonian marls and turbidites

### Stop 3  Rambla Sierra

Walk upstream for about 200m to where the channel takes a sharp right-hand turn. The section on your left shows a major fault affecting ~40m of red, disorganized Serravallian conglomerates, dipping steeply south (Fig. 7.5A,B), forming part of the southern limb of the Serrata del Marchante anticline. Perched unconformably at the top of the section are buff cemented conglomerates, also affected to some extent by the fault. They are part of a Pleistocene fluvial terrace. Throughout the basin there are remnants of two such terraces; these conglomerates are part of the lower of the two. They pre-date the alluvial fan sequences seen east of Tabernas, and therefore are at least mid-Pleistocene in age.

The red Serravallian conglomerates that form the bulk of the section are also exposed in nearby sections at stream level. Note their disorganized and unsorted nature, with a wide range of clast sizes. Further up the rambla there are sections exposing huge boulders (>1m in diameter: *see* also Fig. 3.4A) within this conglomerate. The clasts are of Filabride relatively high grade metamorphics, including numerous garnet micaschist, amphibole micaschist and some tourmaline augen gneiss clasts (*see* chapter 3). Also note the relatively high matrix content between the clasts. The fabrics are compression- and shear-induced fabrics, with little or no indication of fluidity. We interpret the origin as of cohesive debris flows, relatively near source, on steep uplifting terrain, almost certainly of sub-aerial origin.

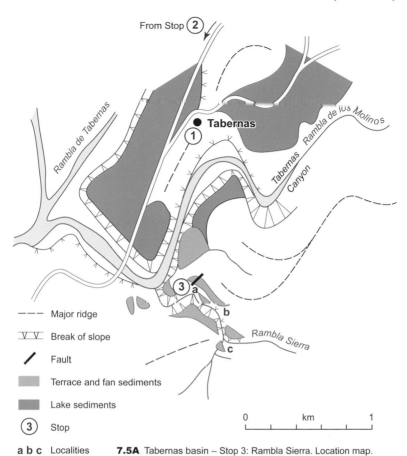

**Legend:**

- – – – Major ridge
- ⊻⊻ Break of slope
- ／ Fault
- (light grey) Terrace and fan sediments
- (dark grey) Lake sediments
- ③ Stop
- **a b c** Localities

**7.5A** Tabernas basin – Stop 3: Rambla Sierra. Location map.

**7.5B** Rambla Sierra fault within Serravallian terrestrial conglomerates. This fault essentially fronts the Serrata del Marchante ridge, the prominent ridge in the south of the basin and south of Tabernas village.

About 200m further up the rambla, exposed on the right (facing upstream) are Pleistocene buff conglomerates that form a terrace within the valley of the rambla. They are set well below the higher level terrace seen on top of the faulted section, and obviously followed a major incisional phase of the modern rambla. Fragments of this terrace can be identified elsewhere along the rambla. Downstream, at the confluence with the Rambla de Tabernas, they seem to have formed a fan into the main valley, partially blocking that valley. This fan, together with a small tributary fan at the exit to Tabernas canyon, may have been responsible for the ponding that created the late Pleistocene Tabernas upper 'lake', at a higher elevation than the lower 'lake'.

Continue along the rambla, following the steeply dipping Serravallian rocks up the sequence. You will notice several trends, involving an increase in the organization of the sediments (*see* Figs 3.4A, and Fig. 7.5C). Sorting improves, clast size decreases and the clasts themselves show greater roundness. The shear and compression fabrics noted earlier are progressively replaced by fabrics suggesting greater fluidity: clast alignment with the flow direction, a clearer separation of clast-rich horizons from matrix. The reddish colour progressively gives way to grey. The changes in fabric suggest not only greater fluidity, but deposition in water as sub-aquatic debris flows. A clear indication comes a little later at ~537996. Barnacle stubs (Fig. 7.5D) are encrusted on a number of clasts and numerous barnacle breccias are distributed through the sediments, together with less

**7.5C** Terrestrial to marine transition (Serravallian to Tortonian). (Photo shows about 3m of section.)

**7.5D** Fossil barnacle stub: clear evidence of marine conditions at the base of the Tortonian sequence.

common oyster and echinoid fragments. Now we are clearly in a marine environment. These trends continue to the next bend in the channel (it turns through a right angle towards the left – east), where another rapid change takes place in the sedimentary environment. There is a rapid transition to sandy turbidites, showing a rhythmic sequence of repeated fining upwards sandstone to siltstone cycles. We are clearly now at the base of the Tortonian turbidite sequence that characterizes the majority of the bedrock geology of the Tabernas basin.

The whole sequence of Serravallian to Tortonian rocks exposed along this section of the Rambla de Sierra represents the initiation of sedimentation in the Tabernas basin, from proximal terrestrial (?) cohesive debris flows to marine deeper water turbidites (for illustration of a similar lithology *see* Fig. 3.4B).

A little general explanation of the Tortonian turbidite-dominated rocks of the Tabernas basin may help to put the features seen at the next three stops into context. There has been a large amount of research reconstructing the submarine palaeogeography of the Tortonian in the Tabernas basin, particularly by Kick Kleverlaan and Peter Haughton. It is impossible here to describe all the sites that contribute to that picture, but we will try to put the Tortonian features that will be seen in the next three stops into their regional context. Sediment was fed into the basin from the north (Filabres) in the form of large submarine fans, involving sandy turbidite sequences of cyclic alternations of occasional gravels, sands, and then muds. Towards the basin centre mud deposition was dominant. An axial channel system developed on the sea floor, orientated W–E, though its destination is not clear. The channels are filled with gravelly and sandy channelled turbidites. Rapid early deepening of the basin led to the dominance of fine sediments, but later sandy and gravelly sheet turbidites became more important. Relatively high in the sequence a major event fed large amounts of fresh, coarse material into the basin, creating a basin-wide megabed, the Gordo megabed, which is interpreted as a seismite, triggered by large earthquake(s) releasing fresh sediment from the surrounding land area. There are other, less extensive and thinner, possible seismites in the upper part of the sequence. At around the same time, submarine slope instability created slump folds in the still poorly consolidated sediment. Stratigraphically above the Gordo megabed is a further sequence of sandy turbidites, with apparently less differentiation across the basin than those below. The whole sequence was folded in late Tortonian times. In the east of the basin the turbidite sequence is unconformably overlain by the uppermost Tortonian Azagador Member. In the west of the basin there is also evidence of an E–W orientated blind fault in the underlying basement that affected depositional patterns during the Tortonian, creating an anticlinal structure, which appears to have continued to be active into the Quaternary, deforming Quaternary travertine and other sedimernts. At this time the Tabernas and Sorbas basins formed one large basin and where not 'independent' basins until the later Messinian.

*Return to your vehicle. Drive back to the main road and turn left (west) and continue for ~1.5km. Park on the road shoulder on a straight section of road before the road takes a bend to the left. This is where you park for stops 4a,b,c (556992) (see Fig. 7.6A). To access Stop 4a (then 4b) walk across a shallow gully, then walk NNW across the level surface of the Pleistocene 'lake' sediments of the Llanos de Rueda, until you reach a track above a deeply incised meander of the Rambla de Tabernas. This is Stop 4a (513994) [W2.4218, N37.0343].*

## Stop 4a  Incised meander of the Rambla de Tabernas at Llanos de Rueda

Beware, this is a hazardous site with a vertical drop of perhaps 20m to the rambla bed below! The cliffs here, created on the outer wall of an incised meander (Fig. 7.6A), expose horizontally bedded sediments of the Pleistocene lower Tabernas 'lake' over dipping Tortonian mudstones

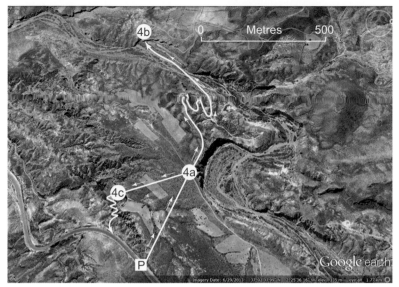

**7.6A** Tabernas basin – Stop 4: Llanos de Rueda (Tortonian rocks, and Pleistocene 'lake' sediments). Google Earth image, showing stop locations and (walking) routes between locations B, C: Tortonian sediments exposed near Site 4b. (North is to the right.)

and turbidites. Within the Pleistocene sediments note the cemented basal gravels, deposited by the river system prior to the ponding that created the lake. They are cemented by a groundwater calcrete where permeability into the underlying Tortonian rocks is impeded. Overlying this are the palustrine and lake muds, followed at the top by fluvial gravels deposited after fluvial drainage had resumed, but prior to the incision that has created the modern incised meander. Note also the youngest (Holocene) fluvial terrace on the inside of the incised meander bend.

*Walk west along the track maybe 200m into the rambla bottom. This is Stop 4b (507993).*

### Stop 4b Sections in Tortonian rocks, showing the Gordo megabed and slump folds: Llanos de Rueda

From where the track crosses the rambla walk downstream maybe 200m (Fig. 7.6A). Exposed, especially on the north side of the channel, is the Gordo megabed (Fig. 7.6B), a bed over 2m thick, an ill-sorted, chaotic conglomerate of relatively fresh schist clasts in a dark matrix of comminuted schist fragments. Exposed in sections on the south side of the rambla are broadly contemporaneous slump folds (Fig. 7.6C).

117

**7.6B** The Gordo Megabed.

**7.6C** Slump folds (within a bed *c.*1 m thick) near site 4b.

*Retrace your steps across the level surface of the Llanos de Rueda almost to the road (Fig. 7.6A). Before you recross the roadside gully, turn to your right to the amphitheatre exposing ~20m of Tabernas 'lake' sediments. This is Stop 4c (513989).*

### Stop 4c  Llanos de Rueda: section in the Pleistocene lower Tabernas 'lake' sediments

About 20m of Pleistocene sediments of the lower Tabernas 'lake' are exposed here (Figs 7.6A, 7.7A). They are predominantly laminated muds, but include sandy horizons, especially near the base and again at the top of the sequence. The sandy layers presumably indicate fluvial input into an otherwise static body of water and mud. There are abundant root remains, suggesting a palustrine (swampy) environment rather than a lake with deep open water, though there are fossil freshwater gastropods and bivalves that indicate aquatic conditions.

We suggest that the 'lake' was created by ponding of the Rambla de Tabernas drainage several kilometres downstream of this site. The 'lake' sediments thin out downstream to almost nothing on the northern flank of the growth fold described in the notes above on the Tortonian rocks (*see* chapter 3).

**7.7** Tabernas lower 'lake' sediments: Llanos de Rueda (Stop 4c). **A**: Section photo (section is *c.*22 m high). **B**: Deformed 'lake' sediments on the upstream side of the growth fold in the lower part of the Tabernas basin, exposed in the motorway cut during construction of the motorway, now unfortunately no longer visible.

Dating the lake sediments has proved problematic. Many years ago, Luis Delgardo obtained a U/Th date by correlation (?) of *c.*150ka on rodent bones extracted from the base of the Tabernas (upper?) Lake sediments. More recently, Martin Geach of Plymouth University has dated the base of the upper lake sediments to *c.*40ka, and we have luminescence dates of *c.*20ka for the top of both upper and lower lake sediments. On this basis all we can say about the initiation of the lower 'lake' sedimentation is that it would have been some time before the date of *c*.40ka on the much thinner sediments at the base of the upper 'lake' sequence.

The dates suggest that for a considerable period during the late Pleistocene drainage of the Tabernas basin was impeded; hence slope erosion would have been reduced. The modern incision wave took place rapidly during the latest Pleistocene as incision into the lake sediments caused local base levels to fall, generating badland slope erosion by incisional undermining of the base of the slopes. [Note that, during motorway construction 2km to the SW beyond Bar Alfaro (Stop 5a), there were superb sections in the late Pleistocene lower Tabernas 'lake' sediments, showing intense deformation (Fig. 7.7B), further evidence of ongoing tectonic activity in this part of the Tabernas basin. Since the completion of motorway construction these sections are no longer accessible].

Looking NW from this site (4c) two flat-topped hills are visible. These hills are capped unconformably by cemented conglomerate sheets, remnants of the Early Pleistocene (?) beginnings of dissection of the basin. Incidentally, this is where David Nash and Roger Smith demonstrated the existence of 'calcrete sandwiches', a mature pedogenic calcrete at the surface and a groundwater calcrete at the interface with the underlying Tortonian turbidites. Look carefully at these hills. The caprock certainly does not slope down-valley; it does not even appear to be horizontal, but may even have a slight dip upvalley to the east. Is this an optical illusion? If not, perhaps this tilting (?) is an indicator of post-early Pleistocene regional tectonism.

*Return to your vehicle, and continue west on the main road for ~3km. Pass another movie set, now called 'Mini Hollywood/Oasys', a western filmset and zoo, a major tourist attraction. [There is something bizarre about working on the geology or geomorphology of this area to the near-constant background of Country and Western music emanating from Mini Hollywood]. Cross the bridge over the Rambla de Tabernas to Bar Alfaro – now part of the motorway service area where the main road from Tabernas meets the Almeria–Guadix motorway. Park here. This is Stop 5a (495973)* [W2.4466, N37.0166].

**Stop 5a  Area to the north and west of Bar Alfaro**
Leave the parking lot on foot to the NW (Fig. 7.8A). Immediately there is a spectacular view of the gullied hill in front of you (Fig. 7.8B). Note the

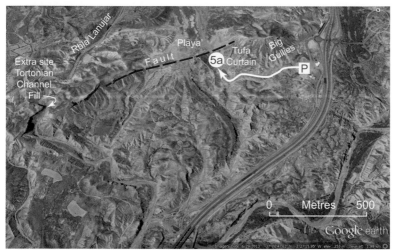

**7.8A** Alfaro area. **A**: Google Earth image of the Alfaro area and area to the west of the motor-way, including (walking) route to Stop 5a (fault and tufa curtains), also showing location of the extra site in the Rambla Lanuja, exposing the Tortonian channel fill.

**7.8B** Big gullies on Alfaro hill (looking north). Note aspect control of small-scale morphology.

aspect control of the modern gullying, with SW-facing slopes almost bare and NE-facing slopes well vegetated (we will discuss gullying processes in relation to Stop 5b, *see* also chapter 5). A fault runs E–W across the face of the hill. This side of the hill there are mudrocks; the hill itself is made of sandy turbidites, becoming sandier upwards.

At the rear of the Petrol station take the small dirt track that descends west into the gully. In front of you is a beautiful tufa curtain, like a frozen waterfall (Fig. 7.8C). The tufa is precipitated by carbonate-rich spring water issuing from the E–W fault that crosses behind the tufa. THIS IS A PROTECTED SITE – DO NOT HAMMER OR ATTEMPT TO REMOVE ANY TRAVERTINE.

Extending west from the tufa curtain is the trace of the fault. A range of travertine morphologies (Fig. 7.8D) are exposed on the valley floor and along the trace of the fault, including barrages, pressure ridges and spring pipes, some of which are still active.

To your north is a flat area encrusted by carbonate salts, creating a playa-like landform. This is the floor of an abandoned course of the Rambla Lanujar (Fig. 7.8E), the drainage about 1km to your west (for afficianados of Indiana Jones, this site featured in The Last Crusade). (We suggest the Rambla Lanujar as one of the other sites in the Tabernas basin that is worth a visit. *See* the last section of this chapter.)

*Return to Bar Alfaro, rejoin your vehicles and drive away from the motor-way intersection back towards Tabernas for about 200m to the far side of the bridge over the Rambla de Tabernas. There is a track on the right of the bridge: park here for the walk to site 5b (El Cautivo badlands). Follow the track onto the floor of the rambla. Follow the rambla bed downstream (south), through a left-hand curve then a right-hand bend. Keep to the left of the channel below an undercut cliff in Tortonian mudrocks. Note the toppled blocks. Where the rambla turns west away from the undercut slope, continue southwest along the edge of the floodplain. After about 150m there is a path up the ridge to your left. Take this path to the top of the ridge. This is Stop 5b (497966)* [W2.4425, N37.0121].

**Stop 5b   El Cautivo badland site**

This is the El Cautivo badland site (Fig. 7.9A). It is an experimental site run by the CSIC (Spanish Government Research: Conseco Superior de Investigaciones Cientificas) to monitor hydrological, geomorphological and ecological processes in dryland environments. Do not walk on the site itself. Do not touch any instruments. You get a perfectly adequate view of the overall geomorphology from here. We have given a full description of the modern processes and interactions between hydrological, geomorphological and ecological processes in chapter 5.

**7.8C** Tufa curtain.

**7.8D** Travertine morphologies – an exposed section through a 'spring pipe'.

**7.8E** Playa and abandoned valleyfloor.

The badlands are cut into Tortonian mudrocks. The early stages of the geomorphic evolution are a series of three gently concave pediment surfaces cut along the divides between the main gully systems of the badlands. The lowest of these pediments appears to grade into the level of the surface of the late Pleistocene 'lake' sediments. It is most likely that the higher

**7.9A** Tabernas badlands: El Cautivo site. Google Earth image showing the (walking) route from Bar Alfaro to the viewpoint overlooking the El Cautivo badland site (*see* also Figs 4.4B and 5.4B for general views of the El Cautivo badlands).

**7.9B** Details of badland surfaces: note aspect control (view towards the NW), and basal slopes mantled with loose debris.

**7.9C** Lichen-covered badland surface.

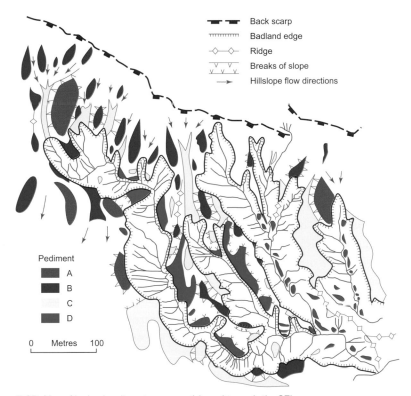

Back scarp
Badland edge
Ridge
Breaks of slope
Hillslope flow directions

Pediment
A
B
C
D

0    Metres    100

**7.9D** Map of badand pediment sequence (viewed towards the SE).

pediments relate to phases of slope erosion and scarp retreat earlier in the Pleistocene. The soil development and vegetation characteristics on these surfaces show a much greater maturity than on the younger badland surfaces. Cut into these surfaces are the main badland gullies. Note the asymmetry of the gully slopes (Fig. 7.9B; *see* also Fig. 5.4B) with bare, steep SW-facing slopes and gentler, partially vegetated NE-facing slopes. The detailed surface processes, studied by Roy Alexander and Adolfo Calvo, differ between these two slope aspects (*see* chapter 5). The SW-facing slopes desiccate more quickly, so they weather more slowly. Infiltration capacity is less, promoting surface runoff and rill erosion. The NE-facing slopes have a greater vegetation cover, including a lichen cover (Fig. 7.9C). After rain these slopes retain more moisture, promoting weathering of the underlying marl, including surface cracking. Therefore infiltration capacity is higher and runoff is less, but sediment removal takes place by shallow mass movement, by mudslides.

It is tempting to assume that this reason for the slope asymmetry is the whole story. However, there may be another, at least partial, explanation. The streams have a tendency to migrate eastwards laterally, undercutting the SW-facing slopes. Furthermore, when the longer term drainage evolution of the badland gully system is reconstructed (Fig 7.9D), there is evidence for several headwater gully captures by streams migrating east. These factors, together with the late Pleistocene uplift of the anticline further downvalley that caused the ponding of the Tabernas 'lake', and the easterly dip (?) of the mesa caps across the valley (described earlier), all suggest the possibility of tectonic sag or relative subsidence of the eastern part of the basin, that is, the part of the basin now occupied by the late Pleistocene alluvial fan systems.

*Return to your vehicle. If you wish to complete the transect of the Tabernas basin we suggest that you continue south to Stop 6. Either use the old road or the motorway; drive for ~6km south to the next motorway exit (for Gador and Rioja). There is another service area here to the east of the roundabouts connecting to the motorway. Park here; this is Stop 6, the final stop on the Tabernas excursion (4778907), [W2.4669, N36.9605].*

### Stop 6  Near Rioja – confluence area of the Rambla de Tabernas with the Rio Andarax

Using the degraded track, scramble up the hill at the back of the service area for an overview of the confluence zone of the Rambla de Tabernas with the Rio Andarax. En route from Bar Alfaro, after about 4km (at 477930) you crossed from the Tortonian rocks onto the Pliocene Gador Formation conglomerates, unconformably cut into the Tortonian rocks, also faulted. The terrain is still highly erosional, but the texture of the terrain on the conglomerates differs from that on the Tortonian rocks. Note also that both the Rambla de Tabernas and the Rio Andarax are channelized here, and the flat valley floors are used for intensive orange cultivation.

To the southeast the Gador conglomerates pass laterally into and overlie sandstones and conglomerates of the Pliocene fan delta, studied by George Postma, within the Rioja corridor. We do not know this area in detail, so cannot recommend specific field sites.

### End of Excursion 2
*******************************

## Other sites within the Tabernas basin that are worth a visit

### The drive along the mountain front of the Sierra de Alhamilla from Lucainena to Turrillas

A drive along a minor road gives spectacular overviews of the geology of the western part of the Sorbas basin and the eastern part of the Tabernas basin. 1:25000 map: Lucainena de las Torres.

*Make your way to Lucainena (713000)* [W2.2001, N 37.0414]. *Leave the village towards the NW on the road towards Sorbas. 200m beyond the village turn left onto a minor road towards Turrillas.*

The road to Turillas hugs the base of the mountain front of the Sierra de Alhamilla. To your left is the fault zone of the northern boundary fault of the Sierra de Alhamilla, with purple Triassic marls in the fault zone itself, south of which are the dark schists of the Sierra de Alhamilla. As you climb towards the divide between the Lucainena drainage and the Tabernas basin there are extensive views to your right (north). Below you, the Lucainena drainage is cut into tightly folded Tortonian sandy turbidites. Beyond, and dipping away from you, is the escarpment of the Azagador Member. Beyond that escarpment, in the middle distance, is the even terrain of the final basin-fill surface (the little-dissected top Gochar surface) of the western end of the Sorbas basin. Beyond, in the distance, is the Sierra de los Filabres.

A good viewpoint is near the watershed between the two drainages (683994). To the right (east) is the aggressive, deeply incised drainage at the head of the Lucainena system, studied by our colleague Elizabeth Whitfield, which over the Pliocene to Pleistocene progressively captured first Sorbas, then Tabernas drainage. (The early stages of this progressive capture sequence are evident in the Gochar-age sediments at La Cumbre in the Sorbas basin, described in the section of this text devoted to 'other localities in the Sorbas Basin that merit a visit' – *see* Chapter 6.6.) Contrasting with this incised terrain are the relatively smooth slopes to the NW (ancient fan or pediment surfaces) at the head of the Tabernas drainage.

As you drive on, the view NW changes (Fig. 7.10). The thrust-forward Marchante anticline brings up the Serravallian conglomerates in front of the Azagador escarpment.

Continue up into Turrillas village by turning left at 661988. Note the landslipped terrain along the Alhamilla boundary fault zone. Turrillas is a

127

**7.10** View north from near Turillas, on the northern margin of the Sierra de Alhamilla, across the eastern end of the Tabernas basin. The Serrata del Marchante ridge in the middle distance is a thrust forward anticline in Serravallian conglomerates. Beyond are the Quaternary Filabres alluvial fans issuing from the stable mountain front of the Sierra de los Filabres.

charming little village perched high on the Alhamilla mountain front. The views from here are even more extensive, especially that to the NW. The gap through the anticlinal Marchante ridge, followed by the Norias drainage, is evident. Beyond that is the smooth terrain of the coalescent alluvial fans of the upper (eastern) part of the Tabernas basin, derived largely from the Sierra de los Filabres.

Leaving Turrillas the way you came (there is no other road), turn left at the road junction, towards Tabernas. The road leaves the mountain-front zone, crossing older alluvial fan or pediment surfaces into the drainage of the Rambla de las Norias, and through the gap in the Marchante ridge. Beyond there the rambla forms a large low-angle, fluvially dominant alluvial fan, wrapping around the much steeper (debris-flow dominated) smaller fans on the front of the Serrata del Marchante. Continue to the main road. There, turn left for Tabernas or right for Sorbas.

### A hike to the steep Quaternary alluvial fans at the mountain front of the Serrata del Marchante

*To access the start of the hike, take the Sorbas–Tabernas main road to a point ~150m east of the Venta del Compadre bar (579036) [W2.3583 N37.0729] (Fig. 7.1). From there take the dirt road to the southwest into the channe of the*

*Rambla de los Molinos. Note the sections in distal fan deposits of the Quaternary Filabres fans (see Tabernas basin excursion Stop 1: interbedded with laminated silts of the uppermost part of the late Pleistocene upper Tabernas 'lake' (see Tabernas excursion Stops 3 and 4c for discussion of the lake sequence). In the rambla the track turns east, then exits the rambla to the south, then swings east again from where you climb out of the rambla. Continue along this track for ~2km to 596030 [W2.3300 N37.0658], where there is a rough track to the right (south). Park near here and continue on foot, heading south up the rough track.*

Ceporro fan, the largest of the Marchante fans, is in front of you (Fig. 7.11A). Keeping an almond plantation (?) to your right, you are soon on the distal surface of the fan. Make your way up the distal fan surface, noting modern lobate cobble bars (Fig. 7.11B), dominantly of garnet-mica schist clasts, derived from the Serravallian conglomerates that constitute the Marchante ridge. Head for the intersection point, the terminus of the

**7.11A** Ceporro fan (for map and profile of this fan *see* Fig. 7.4A,B – contrast with Mezquita fan (visited on Stop 2a). View upfan.

**7.11B** Ceporro fan, distal surface.

**7.11C** Ceporro fan: section exposed in the fanhead trench. Note: debris-flow dominated sedimentary sequence (F1) at the base, overlain by sheet gravels (F2); red soil with calcrete at the top of the section. At this site F2 sediments are cut into F1 sediments. Elsewhere the junction is marked by a buried soil and calcrete horizon.

fanhead trench at 597020 [around W2.3281, N 37.0592]. Now head up the proximal fan surface to the east of the fanhead trench to the fan apex at 600014 [around W2.3266, N 37.0559].

Note the sections exposed across the fanhead trench (Fig. 7.11C). At about 598016 [around W2.3266,N37.0559], basement is exposed on the channel floor and forms a step in the channel profile. This is caused by a thrust fault, part of the system that created the mountain front here. Note the sections. At the base are debris-flow deposits (forming unit F1 of our nomenclature – *see* Tabernas excursion, Stop 2a), including several palaeosols. That unit is part capped by a palaeosol, part truncated by an erosional horizon over which the upper unit (F2) is dominantly of sheet gravels. Unit F2 forms the fan surface, capped by a red soil and a mature calcrete crust. Unit F3 is cut into the fan surface and forms a terrace of unconsolidated gravels, within the fanhead trench.

Note the contrast between this fan and Mezquita fan (Tabernas excursion: Stop 2a; *see* also Fig. 7.4A,B). The sequence of fan deposition and incision differs, and on the faulted Marchante mountain front there is little backfilling of fan deposits into the feeder catchment, in contrast with the fans on the non-faulted Filabres mountain front. *Return to your vehicle the way you came.*

## Rambla Lanujar: excellent exposures of Tortonian turbidites and axial channel fill

*Go to Bar Alfaro Service Station (Stop 5a on the Tabernas basin excursion), but continue on the old road to the service road that parallels the motorway to its NW. Continue to where it descends into the dry river bed. Park in the shade of the motorway flyover. W2.4549,N37.0081. Turn right (NW) along the rambla floor, then after ~100 m turn right again up a large tributary rambla (not the steep ravine). This is Rambla Lanujar – walk upstream.*

The sections in the rambla sides initially show bioturbated marls, then at ~477966 you cross the fault zone seen at Stop 5a on the Tabernas excursion. Beyond the fault you pass into a sandy turbidite sequence. A little further on (location shown on Fig. 7.8A), where the modern channel is incised and constricted, is an excellent exposure of a palaeochannel sequence within the turbidites. Low down in the exposed channel (visible at stream level in the east wall of the rambla, and above a ledge on the west wall) is a bouldery debris-flow conglomerate. The shear and compression fabric indicate palaeoflow to the east [W 2.4640, N 37.0145]. Above this are a series of nested channels, which are ultimately capped by marly sediments. The whole complex has been interpreted by Peter Haughton as part of the basinal axial channel system.

On the sides of the rambla, note the badland development, with microtopography differing in relation to aspect and parent material.

*Return to Bar Alfaro. You can do this by continuing upstream on the rambla floor, then scrambling out of the rambla on the east side along a degraded track at 482977 [W2.4608, N 37.0186]. This will put you on the floor of the abandoned course of the Rambla Lanujar (see Tabernas Excursion – Stop 5a: Fig. 7.8A). Walk across the 'playa' surface to the tufa curtains and scramble back up to Bar Alfaro.*

**8.2** View NW from Mojacar village (Stop 8.1a).

overlain by Tortonian sandy turbidites, then unconformably by the top-Tortonian Azagador calcarenites. The Azagador forms a prominent ridge to the west of Mojacar, culminating in the hill about a kilometre to your NW where it dips steeply to the NW. Beyond the Rio Aguas to the NW lies the subdued terrain developed on the Messinian Abad marl. There are only small patches of age-equivalent Cantera reef in the far western part of the basin.

From this point in the sequence the stratigraphy diverges from that of the Sorbas basin. There is no Yesares Member massive gypsum exposed in the Vera basin, but the uppermost Messinian (?) marls that crop out in the basin centre are riddled with gypsum veins. Overlying these marls are yellow bioturbated sandstones of the Pliocene Cuevas Member, visible as the yellowish terrain in the basin centre about 10km to your north. The Pliocene marine sequence of the Vera basin culminates with the fan-delta conglomerates of the Espiritu Santo Formation (*see* also Fig. 3.9), which form the hills in the basin centre 10–15km north of you. The basin filling sequence is followed by the Plio-Pleistocene alluvial-fan and fluvial conglomerates of the Salmeron Formation, which form the greyish terrain in the far NW of the basin about 15km to your NW.

The other features visible from here relate to the general structural and topographic setting. The basin is truncated to the east by the coast-parallel Palomares fault system. For about 12km towards the north the fault system lies along the coast, but beyond there it lies to the west of the Sierra Almagrera (the mountains on the coast *c.*15km to your NNE), along the

Pulpi corridor. To the west of the Pulpi corridor is the Sierra Almagrera, bounding the Vera basin to the NNW. The western boundary of the basin is formed by the eastern part of the Sierra de los Filabres, including the metagranite of the Sierra de Bedar and the metacarbonate of the Sierra de Lisbona.

*Return to your vehicle(s) and leave Mojacar by heading down the hill, then turning west towards Turre. After about 1km, on a right-hand bend there is a track on the left (at 013114 [W1.8650, N37.1450]). Turn into this track and park [W1.8682, N37.1395]. On the south side of the track there is a stream channel. Walk the c.50m WNW to this channel to see an enormous collapsed pipe – Stop 1b.*

### Stop 8.1b Collapsed pipe, west of Mojacar

This is perhaps a curiosity, but it is an interesting site (Fig. 8.3), and worth a brief stop. We do not know of any larger collapsed pipe – and on such a small drainage. The pipe itself is in Quaternary gypsiferous sediments reworked from the Tortonian sandy rocks within the catchment. The roof of the pipe is highly indurated with carbonate. The origin of the pipe is not clear, but is likely to be related to the different hydrological properties of the underlying Tortonian rock and the overlying Quaternary sediments, and may in part owe its origin to historic irrigation practices.

**8.3** Large pipe developed west of Mojacar (Stop 8.1b).

*If not visiting the quarry stay on the road for another 300m. Park here, on the right. This is Stop 8.4d. (If you visited the quarry, retrace the track back to the road and turn left. After 300m park on the right at Stop 8.4d [W1.8959, N37.2846].)*

## Stop 8.4d  Oyster bed within the brightly coloured Pliocene Cuevas Formation sandstones

This site justifies a brief stop, not only to see the brightly coloured sandstones within the Pliocene Cuevas Formation, but also to examine a richly fossiliferous oyster-bed horizon within that formation (Fig. 8.9). Within this Pliocene shallow-marine sandstone, the section is dominated by large ostrea fossils, from large colonies that lived in the vicinity of the fan delta, before slumping downslope.

**8.9** West of Cuevas de Almanzora. Stop 8.4d: Oyster bed section in Pliocene Cuevas Formation.

*Return to your vehicle(s). Continue west along the road for c. 2.5 km to the motorway intersection. There turn south onto the old N340 for another c. 1 km to a roundabout. Take the second exit from the roundabout onto the old road. this road meanders right then left then right again. On this bend (about 400 m from the roundabout), drive under the new highway bridge to a point above the confluence of the Rambla Cajete with the Rio Antas, just short of the village of Antas. This is Stop 8.5a  (960234)[W1.9170, N37.2506].*

## Stop 8.5a  Rio Antas

The Rio Antas drains a large basin from within the Sierra de los Filabres on the western side of the Vera basin. At Antas it is joined by the Rambla del Cajete, which drains the northwest segment of the Vera basin between the Sierra Almagrera and the Sierra Lisbona, the eastern extremity of the Filabres. The Antas is the much larger stream. In the past, during the Pleistocene, the Cajete did not join the Antas but continued eastwards through the broad, open valley to the north of Vera. At some stage, presumably during the late Pleistocene, although we do not know the details, a steep tributary of the Antas (which was at a lower elevation than the Cajete) cut back and captured the Cajete (Fig. 8.10). There is a steepening in the profile of the modern channel of the Cajete about 500m north of Antas village that represents the capture knickpoint.

*Return to your vehicle(s). Drive across the Rio Antas into Antas village. The road through the village makes a sharp left-hand bend, about 100m after which you turn right onto a minor road heading for Jauro. Follow this road for about 3.5km. Note the extensive citrus orchards in the Antas valley to your right; note*

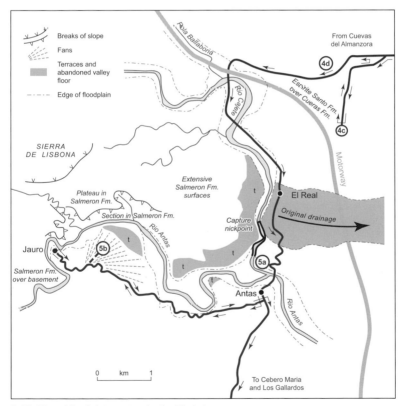

**8.10** Map of the Cajete capture by the Rio Antas.

*also the Roman(?) aqueduct across the river. After about 3.5km the road begins to climb steeply up a hill, and through a roadcut in grey conglomerates. Park on the shoulder immediately after the roadcut. Walk about 250m onto the hilltop to the north of the road, where you get an unobstructed view across the Antas valley to the north (928235)* [W1.9540, N37.2506]. *This is Stop 5b.*

## Stop 8.5b East of Jauro village

In front of you the Rio Antas cuts a large section in the Plio-Pleistocene Salmeron conglomerates (Fig. 8.11). These are terrestrial conglomerates, and followed deposition of the upper Pliocene Espiritu Santo fan deltas, seen earlier. They represent the switch to a continental environment after the marine environment that characterized the area for most of the Neogene. The section shows two bodies of sediments. At the base are reddish conglomerates with occasional sandstones, with a clast content of metacarbonates derived from the Sierra Lisbona (the range visible to the north of the section). Our colleague Martin Stokes has interpreted these sediments as alluvial fan sediments fed into a local fan from the Sierra Lisbona. They are overlain by more mature grey conglomerates, representing the braided river environment of the palaeo-Rio Antas. Clast content in these sediments includes Filabride schists and tourmaline gneiss, clasts from the Sierra de Bedar to the west. The whole section is capped by a massive pedogenic calcrete. Both pedogenic and groundwater calcretes

**8.11** Plio-Pleistocene Salmeron Formation conglomerates east of Jauro village. Alluvial fan deposits at the base, fluvial gravels above.

are exceptionally well developed in the zone around the Sierra Lisbona because of the availability of carbonate from the metacarbonate rocks of the Sierra Lisbona.

In this mountain-front zone two Plio-Pleistocene alluvial fans were present: the Calentones fan, fed from the Sierra Lisbona, and the Loma del Perro fan, fed from the Sierra de Bedar to the south, and underlying the ground on which you are standing. Both fans were succeeded by a braided river system as the drainage became more integrated.

There are sections in fan and braided stream sediments exposed in stream sections in the vicinity of Jauro village, which also show major post-depositional tectonic disturbance. To visit these sections continue to Jauro village and turn right down a track into the river bed. For sections exhibiting details of fan sedimentology, turn right for about 200m (downstream). For evidence of tectonic disturbance, turn left from the village, follow the track into the river bed and walk 400m (upstream).

Note: it is advisable to park before the village, as Jauro village is impossible for any vehicle larger than a very small car.

*Transfer to Stop 8.6. Drive back fom Jauro village to Antas. In Antas turn right (south) onto the road towards Los Gallardos. After about 4km turn right onto a minor road towards the Ermita de la Virgin de Cabezo Maria (the shrine on top of the black volcanic hill to your west). The road skirts the hill to the south, then feeds into a parking area (946194) [W1.9351, N37.2109], from where you can climb to the Hermitage. This is Stop 8.6.*

## Stop 8.6  Cabezo Maria

In addition to the Neogene sedimentary basin-filling rocks, there are volcanic rocks present in both the Vera and the Almeria basins. These are not directly related to the Cabo de Gata volcanic suite (*see* chapter 10). They are part of a group of Tortonian to Messinian rocks associated with volcanic activity to the west of the Carboneras and Palomares fault systems. Cabeza Maria in the Vera basin is a Messinian volcanic centre, comprising today an eroded volcanic neck (Fig. 8.12) from which lava flows emanated. There are remnant patches of the lava flows elsewhere in the centre of the Vera basin. The centre of the Cabezo Maria volcanic mass is composed of dark potassium-rich lavas, also rich in olivine. Towards the margins of the volcanic centre brecciation with the country rock is evident, as is interbedding of the lavas with the country Messinian rocks.

**8.12** Cabezo Maria volcanic neck.

Walk to the top of the hill, not only to inspect the volcanic rocks, but for the panoramic view over the western part of the Vera basin.

*Return to your vehicle(s). Drive back to the main road, turn right (south) for the 2.5km to the motorway intersection north of Los Gallardos.*

### *End of Excursion 3*
*******************************

# Chapter 9

# Sierra Cabrera, the coast and the Alias valley (Excursion 4)

## Highlights

Intersecting fault systems exposed along the coast; the Pleistocene coastal sequence; drainage adjustment to the fault systems.

The route skirts the Sierra Cabrera, starting on the margins of the Vera basin, then follows the Palomares faults along the coast (which also exhibits excellent exposures in the Quaternary coastal sequence). Several stops are made to examine the influence of base-level change and of the Carboneras faults on the drainage evolution of the Rio Alias. Finally, we examine the transverse origins of the Alias drainage in the Feos valley (the link from the Sorbas basin drainage, examined on Day 1) and in the Polopos area (Fig. 9.1).

Key locations: Mojacar, Macenas, Sopalmo/Sierra Cabrera, Carboneras, Llano del Antonio, Argamasson, Feos Valley, Polopos.

We will include a brief description of each location visited.

## The coastal area south of Mojacar

*Start at the road junction on the sea front at Mojacar, where the road from Mojacar village joins the coast road (041107)* [W1.8300, N37.1348]. *Drive south along the sea front/coast road, past the Hotel Indalo, where the road turns inland. Note the recent construction of groyne-like structures built to 'improve' the beach facilities. After a couple of kilometres the road again runs alongside the sea. At the roundabout at the mouth of the Rambla Macenas, turn left onto a gravel road that first crosses the modern fan delta of the Rambla de Macenas, (note the Napoleonic(?) tower on your left). The track then hugs the shore on low cliffs. After about 1 km from the roundabout there is space on the right to park. Park here. This is Stop 1: Macenas beach (024035)* [W1.8503, N37.0715].

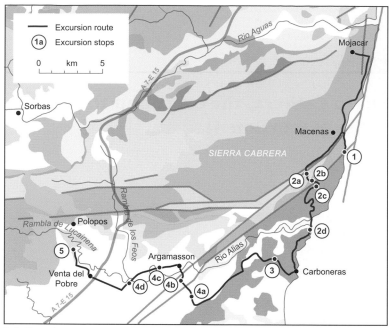

**9.1** Sierra Cabrera/Rio Alias: geological map, excursion route and stop locations for Excursion 4. For key to underlying geology *see* key on p. 64.

## Stop 9.1  Macenas Beach

Macenas beach lies at the intersection of the Palomares and Carboneras fault systems (Fig. 9.2A). The coast (aligned more or less N–S) parallels the Palomares fault system, currently the more active of the two systems. The Rambla de Macenas, together with the parallel valley that meets the coast at the parking place, are aligned with two strands of the (NE–SW orien-tated) Carboneras fault system. The bedrock involved here is dominantly dark Cabrera schist, but incorporated within the second fault (from the north) of the Carboneras system are fault rocks, Triassic marls of variegated colours (visible at the parking place). A little to the south of the parking place, where the two fault zones intersect, are what have been interpreted to be Oligocene limestones transported in along the Palomares faults.

Unconformably resting on the basement rocks is a superb sequence of Pleistocene shoreline conglomerates (Fig. 9.2B,C; *see* also x on Fig. 4.2A; see also Fig. 4.2B). They rest on an irregular erosional base at and below modern sea level, above which are ~5m of cemented, well sorted, well rounded, quartz-dominated beach conglomerates. Occasional fossil

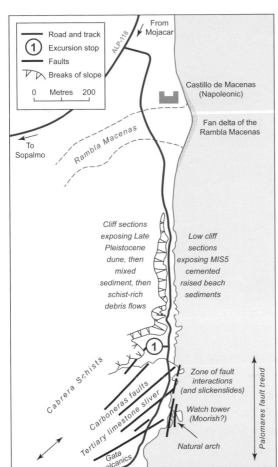

**9.2A** Macenas coast (*see* Fig. 4.2A for an interpretive section; Figs 4.2 B,C for other photographs of this site; and Figs 5.6 A,B for sequential photographs showing erosional change).

Map labels:
- Road and track
- Excursion stop
- Faults
- Breaks of slope
- 0 Metres 200
- From Mojacar
- ALP-118
- Castillo de Macenas (Napoleonic)
- Fan delta of the Rambla Macenas
- Rambla Macenas
- To Sopalmo
- Cliff sections exposing Late Pleistocene dune, then mixed sediment, then schist-rich debris flows
- Low cliff sections exposing MIS5 cemented raised beach sediments
- Cabrera Schists
- Carboneras faults
- Tertiary limestone sliver
- Gata volcanics
- Zone of fault interactions (and slickenslides)
- Watch tower (Moorish?)
- Natural arch
- Palomares fault trend

**9.2B** Coastal sedimentary sequence (see Figure 4.2A for interpretation): coastal sediments.

**9.2C** Coastal sediments: note buried cantilever structure.

bivalves and gastropods are present at the top of the sequence, including the gastropod *strombus bubonius* (*see* Fig. 4.2C), part of the subtropical Senegalese fauna, characteristic of full Pleistocene interglacials, but not present in the modern Mediterranean, nor since the last interglacial. These sediments almost certainly relate to the last interglacial (Tyrrhenian III, MIS 5, 125ka; *see* Appendix 3).

Interestingly, there are two erosional horizons within the sequence, marked by toppled cantilever structures (Fig. 9.2C), wrapped around by a fresh basal conglomerate followed by more beach conglomerates. We interpret this sequence to represent a cyclic progressive transgressive sequence, whereby during rising sea level, subaqueous cementation of the first beach gravels took place (our colleague Jim Marshall examined the cements petrographically and concluded that these are typical of a marine environment). Sea level then fell slightly, exposing the now cemented rocks to erosion as sea level rose again. Erosion of the low cliff produced the cantilever failures, which then became buried, first by the new basal conglomerate, then by more beach gravels. This sequence was repeated, suggesting an oscillatory rise in sea level to a maximum ~5m above the modern sea level.

The beach conglomerates are overlain by sands that show low-angle cross bedding near the base (y on Fig. 4.2A). They also show tubular structures suggestive of root casts. Further up the sequence they are structureless, and higher up the slope become dirty, mixed with soil and terrestrial debris. They are overlain by debris-flow deposits derived from the schist slopes above (z on Fig. 4.2A). We interpret the origin of the sands to be aeolian dunes that were plastered over the cliffs (Fig. 9.2D), probably during marine regression, which exposed a foreshore as a potential sediment source. Later the upper parts of the plastered dunes became unstable and collapsed down the slope, incorporating slope material within the reworked sand. Finally, probably during the ensuing 'glacial', increased weathering and slope erosion produced debris flows that buried the sequence (*see* also Fig. 4.2A,B). All these features are visible within 100m of the parking place.

After examining the Pleistocene sequence, walk south along the track. At the first rock outcrop examine the faults evident on both sides of the track, affecting the limestone rocks (apparently of Oligocene age). On the landward side of the track is a minor fault with a NE–SW orientation (Carboneras fault trend). Slickensides (Fig. 9.2E) indicate dominantly

**9.2D** Aeolian sands – above the beach sediments, overlain by terrestrial debris-flow deposits.

**9.2E** Palomares fault-trend slickensides in Early Tertiary limestones, incorporated into the fault zone.

lateral movement. On the seaward side of the track is another minor fault, but with a N–S alignment (Palomares trend, truncating the Carboneras trend), again with dominantly lateral movement.

Note also the limestone blocks on the foreshore. At water level they are cut by what appear to be wave-cut notches. However, some biological activity is involved. Careful inspection of the notches reveals borings lined with green algal matter. It is likely that acid-secreting algae are, at least in part, responsible for the notch formation (see chapter 5, section 5.4, p. 59).

Continue for ~100m along the track, until you are above a headland, crowned by a watchtower (of Arabic age?). Take the path down, then up to the base of the tower. Take care – there are vertical drops onto rocks. You cross two (N–S orientated, Palomares trend) faults, separating the tower 'island' from the mainland.

Take in the view from the base of the tower. Note the beach immediately to the north. In the 1980s this was a continuous sand beach below the low cliffs in the Pleistocene shoreline sediments. Since then the sand has disappeared, leaving only cobbles and rock (see Figs 5.6A,B). We suspect that the cause is marina and groyne construction at Mojacar, trapping sand that would otherwise be free to move south by longshore drift. Hence the sediment starvation here at Macenas beach.

Now look further north. You are looking along the alignment of the Palomares fault system. To its left (west) is the truncated Sierra Cabrera. In the far distance, to the right (east) of the fault line is the Sierra Almagrera, formerly the eastward extension of the Sierra Cabrera, that has been tectonically displaced north by ~15km along the Palomares fault, according to Martin Stokes, since the Miocene. In doing so it has opened up the Vera basin to the sea (see chapter 8, Excursion 3). Looking south, to the west of the Palomares fault that more or less parallels the coastline, is terrain affected by the Carboneras fault system. First there is terrain on the Cabrera schists, then, beyond the southern fault of the Carboneras system, are the Miocene volcanic rocks of the northern part of the Cabo de Gata volcanic province, here brought up from the south against the basement Cabrera schists.

Go back to the track. While here, it is worth walking another ~100m south along the track to see the modern erosional coast. There is a natural arch in a coast-parallel ridge (see Fig. 9.2A), beyond a trough that picks out a Palomares fault.

*Return to your vehicle and drive back northwards to the road. Turn left there towards Sopalmo and Carboneras. The road follows the Rambla Macenas, which is aligned with the northernmost fault of the Carboneras fault system. After about 2.5 km the road crosses the rambla and climbs to the col at Sopalmo. At the top of the col park on the shoulder on the right-hand side of the road. This is Stop 2a: Sopalmo (996020)* [W1.8811, N37.0569].

## Sierra Cabrera

### Stop 9.2a  Sopalmo

The view from the col at Sopalmo allows you to appreciate the geology of the Carboneras fault zone (Fig. 9.3A). There are three main fault strands, the two northerly strands primarily affecting the Cabrera schists. The most northerly, visible to the north of the col, is the one you followed along the

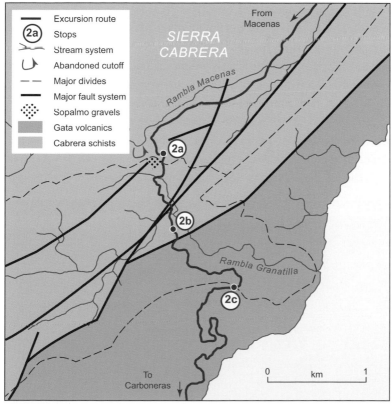

**9.3A** Sopalmo col. Map showing locations of Stops 2a, 2b in relation to the geology and geomorphology.

Macenas valley. The second and third strands are the south of the col. The second will be visible from Stop 9.2b, when you drive south from here. The third, and most southerly strand, is visible to the SW of the col, separating Cabrera schist terrain from the Cabo de Gata volcanic terrain, which forms the large buff-coloured hill to your southwest.

The geomorphology is also fascinating, and presents somewhat of a puzzle. Perched in the col (to the west of the road) are Quaternary fluvial gravels (Fig. 9.3A,B) There has quite clearly been a river capture here, but the evidence relating to which stream captured which forms a puzzle. To the north is the fault-aligned Macenas; to the south the steep drainage of the Granatilla (Fig. 9.3A). The fault-aligned Macenas could well be the subsequent stream that worked back along the fault, capturing the head-waters of the proto-Granatilla, at that time with a catchment to the NW of the col. Alternatively, the steeper Granatilla may have been more aggressive, despite crossing the structural trend, capturing the headwaters of the proto-Macenas, then with a catchment to the SW of the col. The evidence should be preserved in the gravels perched in the col (Fig. 9.3B). The clast alignment of the imbricated clasts within the sediments suggest a palaeo-current from south to north, in other words from the upper Granatilla into the Macenas, indicating that it was the lower Granatilla that was the captor stream. However, during incision, tortuous incised meanders developed, which throw some doubt on the validity of that interpretation of the evidence. Furthermore, if that were the case, there ought to be a provenance signal within the sediments. The modern upper catchment of the Macenas

**9.3B** Quaternary fluvial gravels perched in Sopalmo col – from which direction were they deposited?

is wholly on Cabrera schists, but the Granatilla drains a catchment at least 25% on the volcanics, the rest on Cabrera schists. If the evidence suggested by the palaeocurrents is correct, then we would expect a reasonable representation of the volcanics within the clast content of the gravels. We have found very few volcanic clasts within the gravels, and even for those, we cannot be sure that they are *in situ*. So the evidence is unclear. The site remains a puzzle.

*Return to your vehicle and drive south into the valley of the Granatilla. Cross the bridge over the rambla and after ~50m park in an abandoned section of road on the left. This is Stop 9.2b: Rambla de Granatilla (997013)* [W1.8801, N37.0511].

## Stop 9.2b Rambla de Granatilla

The purpose of this stop is to view the second of the Carboneras faults. You have just crossed this fault, but the fault zone, including variegated, colourful fault rocks (Triassic marls) is beautifully displayed across the rambla to your northeast (Fig. 9.3C). The marls have been eroded into a badland terrain to produce 'painted badlands'. The trace of the fault can be followed NE along a small valley up the hillslope. It is possible to descend on foot into the rambla bed here for a 'hands on' exploration of the fault zone. It is also possible to follow the rambla (on foot – or in dry weather, in

**9.3C** 'Painted badlands' developed in variegated Triassic fault rocks, along one strand of the Carboneras fault system, Sopalmo.

a robust vehicle!) to the coast. There are excellent sections in the volcanic rocks (lavas and agglomerates). In addition there are well-developed tafoni (*see* chapter 5; *see* also Fig 5.1), developed in case-hardened volcanic rocks.

*Return to your Vehicle. Continue to drive south, climbing up the south side of the Granatilla valley, across the southernmost of the Carboneras faults onto the volcanic rocks, and up to another col. There is ample space to park on the left of the road. Park here; this is Stop 2c: Col de Granatilla (004006)* [W1.8722, N37.0444].

## Stop 9.2c  Col de Granatilla

There are extensive views north and south from the col. To the north the view is towards Sopalmo, over the terrain you have already passed through. To the south there are extensive views towards Carboneras. The higher ground, extending over the first few kilometres, is on the Miocene Cabo de Gata volcanic rocks. Beyond that, forming the plateau south of Carboneras, are Messinian limestones. The volcanics exposed in the road-cuts at this site are hornblende andesites (*see* Fig. 3.1), a common lithology within the Cabo de Gata volcanic rocks. Note the large brown hornblende phenocrysts.

*Return to your vehicle. Continue to drive south, down the very tortuous hairpin bends, eventually past a new resort development on the left, to the crossing of the Rio Alias/Carboneras, a wide gravel-bed braided river. Park on the river bed. This is Stop 2d: mouth of the Rio Alias (996984)* [W1.8810, N37.0245].

## Stop 9.2d  Mouth of the Rio Alias

Walk east, along the dry river bed towards the sea (<100m), then when above the beach zone turn south for about 100m. There are fragments of a Pleistocene raised beach (cemented quartz conglomerates again, as at Macenas beach) up to ~5m above modern sea level (*see* above). Turn with your back to the sea and walk up the slope. There are remnants of a higher raised beach about 30m above modern sea level. We suspect that this relates to an earlier interglacial than the <5m raised beach (Fig. 9.3D).

*Return to your vehicle. Drive south to Carboneras. At the south end of the town, note further low cliffs exposing Pleistocene raised beach deposits. At the south end of the town itself, turn right onto what was the old road towards El Saltador and El Llano de Don Antonio. Follow the road out of town for ~1km to*

**9.3D** Pleistocene raised beach, near the mouth of the Rio Alias, north of Carboneras (Stop 2d).

*where the road enters and follows a large abandoned incised meander of the Rio Alias. This is Stop 9.3a: east of El Llano de Don Antonio (975958)* [W1.9026, N37.0036]. *Alternatively, if you wish to omit this stop, continue right through Carboneras to the new main road towards Venta del Pobre. Turn right onto this road and continue past the El Llano de Don Antonio for about 8km to the turn on the right for Argamasson. Turn right there towards Argamasson (see itinerary below: Stop 9.4a).*

## The Alias Valley

From Carboneras the excursion route takes us inland into the zone where the Carboneras fault system and the northern and southern Sierra Cabrera fault systems merge. The geology of the area is complex, involving a wide range of rocks outcropping south of the basement schists of the Sierra Cabrera, ranging in age from the Tortonian-age volcanic rocks of the Cabo de Gata suite to Messinian to Pliocene sedimentary rocks. The area is drained by the Rio Alias, studied by Elizabeth Whitfield. That river system demonstrates the effects of the interplay between the various controls over the evolution of fluvial systems: tectonics, climate, base level, river capture. The terrace sequence is similar to that in the Sorbas basin (*see* chapter 6, Excursion 1), with terraces A, B, C, D all present. As elsewhere,

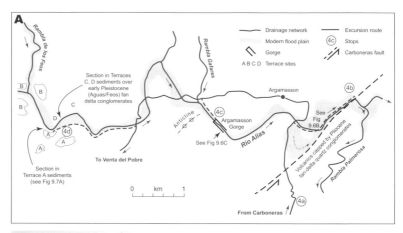

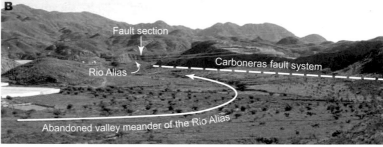

**9.5** Argamasson. **A**: Map of the Alias valley from the crossing of the Carboneras faults, downstream of Argamasson, upstream to the Feos/Alias confluence, including the locations of Stops 4b, 4c, 4d. **B**: Photo of the abandoned meander upstream of where the Rio Alias crosses the Carboneras fault: Argamasson (Stop 4b). **C**: Argamasson canyon knickpoint (Stop 4c), cut within Pliocene sandstones. Note that this gorge not only forms a knickpoint in the modern channel, but also formed a similar knickpoint in the channel prior to the aggradation of Terrace D. Our colleague Elizabeth Whitfield interprets the incisions that migrated upstream as far as the nickpoint as responses to movement on the Carboneras fault.

The skyline to the south, SE of the fault system, preserves the clinoforms of the Pliocene fan-delta sediments seen at the roadside section on the way into Argamasson. They crop out only south of the Carboneras faults. Brought in from the south along the fault system are the Cabo de Gata volcanics, visible immediately across the river (north) from the bridge, and across the main fault ~300m downstream at a bend in the river (Fig. 9.5B). Immediately downstream from the bridge, cemented fluvial conglomerates (possibly part of a Pleistocene river terrace) are exposed on the river bed, truncated by one of the branches of the fault system. Perched a little above the river to the south is an abandoned meander loop (Fig. 9.5 A,B) that runs behind the hill to your southeast, and has since been cut off by the modern channel alignment. Cutoff incised meanders may signify rapid incision. Remnants of the Pleistocene river terrace sequence are preserved here, a relatively young terrace (Terrace D) having been clearly affected by faulting.

*If you have time, it is worthwhile walking downstream to where the modern channel intercepts a main branch of the fault system (Stop 9.4b on Fig 9.5A). Rejoin your vehicle and continue north, then west through Argamasson, over a low divide to where the road crosses the Rambla Gafares. Park here. This is Stop 9.4c: (955953) Argamasson gorge* [W1.9944, N36.9976].

## Stop 9.4c  Argamasson Gorge

Walk SE from the parking place over a low hill to where you emerge above the Rio Alias. Here the Rio Alias is in a canyon, cut into Pliocene coarse cross-bedded shallow marine sandstones. The canyon forms a knickpoint not only in the modern stream profile (Fig. 9.5C), but also in the Terrace D profile. Cemented terrace D gravels are preserved in the canyon walls at the downstream end of the canyon. It is possible to reconstruct the stream profile that preceded the deposition of Terrace D to a knickpoint in the same position as the modern knickpoint. The knickpoints are related to local base-level changes working their way upstream from the fault zone, held up by the resistant Pliocene sandstone. This accords with ongoing movement on the fault system, movement that in the late Pleistocene has been primarily vertical, downdropping to the east.

*Return to your vehicle. Drive on west along the minor road. You cross the channel of the Aguas three times (the river takes a big bend to the north). Park at the third river crossing (874947). This is Stop 4c downstream of the the Feos/ Alias confluence zone* [W2.0195, N36.9930].

### Stop 9.4d Feos/Alias confluence zone

Walk west ~400m from the parking place along the south bank of the river to the top of a prominent hill (Pliocene Cuevas Fm. bioturbated soft sandstone below, capped by terrace gravels of Terrace A, culminating in a mature pedogenic calcrete crust). From the top of this hill there is an excellent view over the Feos–Alias confluence area (Fig. 9.5A).

The Aguas/Feos, draining most of the Sorbas basin (*see* chapter 6) was the main headwater of the Alias until Terrace C time (~70ka). Its headwaters were captured by the lower Aguas and diverted away from this area of the Almeria basin eastwards into the Vera basin (*see* chapter 6; Excursion 1). The diminutive beheaded misfit stream, the Feos, albeit a transverse drainage crossing the basement sliver that links the Sierras Cabrera and Alhamilla, then became merely a small north bank tributary of the Alias, which had lost about 70% of its previous drainage area. Further upstream, the Alias itself is also a transverse drainage across the eastern part of the Sierra de Alhamilla. It heads in the Lucainena area in the SW of the Sorbas basin (*see* below and chapters 6, 7).

From this viewpoint the main terrace sequence of the Aguas/Feos and Alias can be identified. The viewpoint itself is on Terrace A (Fig. 9.6A). It is difficult (hazardous!) to examine the sediments themselves at this site, but there are fallen blocks on the stream bed from which some idea of the sediment provenance may be derived. These contain not only Alpujarride dark schists from the Sierra Cabrera, but also higher grade hornblende schists from the Filabres to the north of the Sorbas basin, demonstrating quite clearly the Aguas/Feos origin of these sediments. Fragments of Terrace A are present further up the Feos valley and to the west in the Polopos area (*see* Stop 5). The younger terrace sequence can be seen below you (Fig. 9.5A). Terrace B forms the flat hill-top terrace surfaces to the north. Terrace C is the calcreted terrace surface below to the east, affected by faulting (Fig. 9.6B). Below that, in the foreground on the inside of the river bend, the extensive lower terrace is Terrace D.

Elizabeth Whitfield has studied these terrace sediments and has demonstrated not only a major change in sediment provenance between Terraces C and D, but also a major change in palaeo-channel geometry, from a large river to a much smaller stream, coincident with the river capture and the loss of the Aguas/Feos headwaters.

**9.6** Near the Feos/Alias confluence (Stop 4d). **A**: Gypsum at the base of the section, overlain by white marls, then unconformably capped by gravels of Terrace A. These gravels include Filabride hornblende schist clasts, therefore were fed by the proto-Aguas/Feos river system from the Sorbas basin. The section is capped by a magnificent indurated calcrete. **B**: Terrace C – faulted and deformed.

Climb down from the hill into the river channel. The walk back to the parking place allows inspection of the sediments of Terrace D and Terrace C.

The first set of sediments, exposed on the left of the channel, belong to Terrace D. There is only one topographic Terrace D here (about 4–5m above river level) as opposed to two terraces more generally elsewhere (labelled D1 and D3). However, careful inspection of the section reveals a major erosional unconformity halfway up the section. Elizabeth Whitfield and Barbara Mauz have obtained OSL dates from the sediments below and above this unconformity, yielding dates of ~30ka and ~15ka respectively, equivalent to Terraces D1 and D3 respectively. Terrace D3 sediments therefore bury D1 sediments with little or no incision between. Both sets of sediments themselves are simple imbricated fluvial gravels, exhibiting shallow channelization and low relief bar forms, typical of the post-capture Rio Alias.

Further along the section, highly indurated conglomerates are exposed just above the stream bed. Their sedimentology suggests deposition by wet debris flows ('swirling' structures with some parallelism of clast alignment). Significantly they are rich in Filabride hornblende schists. They clearly pre-date all the terrace sediments. We interpret these as derived from the Sorbas basin by the proto-Aguas/Feos fluvial system, being fed into a fan-delta environment on the northern margins of the Almeria basin. We suggest that they are the distal equivalents of the Plio-Pleistocene Gochar Formation of the Sorbas basin.

The third exposure shows 7m of Terrace C gravels (Fig 9.6B), with a base eroded into the above-mentioned conglomerates at about river level. These gravels are markedly different from those of Terrace D seen earlier. There are more large clasts. The palaeochannelling is deeper and the palaeo-bar forms are larger, all characteristics of the pre-capture Aguas/Feos system. Also, not seen in the section of Terrace D deposits, there is faulting affecting these terrace sediments. Although we are a little distance away from the Carboneras fault system, we are still in the zone where the Carboneras faults interact with the Sierra Cabrera southern boundary faults. The terrace is capped by a moderately developed pedogenic calcrete, characteristic of Terrace C. Of significance is that the bases of both Terraces C and D are at similar elevations, indicating only limited post-capture incision in this part of the Alias valley.

*Return to your vehicle. Drive west along the minor road to join the Carboneras–Venta del Pobre road and continue into Venta del Pobre. From there take another minor road (north out of Venta del Pobre alongside the motorway on its west side, then turn left – west: towards Polopos. Continue for about 2.5km to a place called Polopillos, where the road descends into the rambla. Park just before the road leaves the rambla to the north (right). This is Stop 5: (824970) Polopillos [W2.0742, N37.0143].*

## Stop 9.5 Polopillos

The rambla is the Rambla de Lucainena, the main modern headwater of the Rio Alias. The Lucainena is a transverse drainage flowing from the SW corner of the Sorbas basin across the faulted terrain of the eastern Sierra de Alhamilla into the NW corner of the Almeria basin. Its origins are probably similar to those of the Aguas/Feos, initially as a superimposed stream from the receding Plio-Pleistocene sea. It was probably superimposed onto the basement rocks of the eastern Sierra de Alhamilla, but with an element of antecedence as uplift of the Alhamilla took place. Today it is incised in a canyon through the Alhamillas (*see* chapter 6, Sorbas basin – end pieces; *see* also Fig. 5.5A). It has extended its headwaters into the SW part of the Sorbas basin in the Lucainena area as a subsequent stream, cutting back through weak Tortonian marly turbidite rocks more or less along the northern mountain front of the Sierra de Alhamilla (*see* also Fig. 6.12A). In doing so, it has extended its drainage area by capturing the headwaters of north-flowing Sorbas basin drainage (for more details *see* chapters 6 and 7).

The section across the rambla exposes an interesting suite of rocks, sediments and structures (Fig. 9.7). At the base of the section are soft, yellow Pliocene bioturbated sandstones (equivalent of the Cuevas Formation of the Vera basin; *see* chapter 8). Cut into these from the north (left) are lobate masses of indurated, ill-sorted conglomerates. These show characteristic fabrics of wet debris flows. Above these are better-sorted bedded conglomerates, in the upper part of which is a well developed palaeosol. This whole suite has been gently folded and is unconformably overlain by more bedded conglomerates and palaeosols. Further along the section towards the east the lower bedded conglomerates and palaeosols are affected by faulting.

Our interpretation is that the sequence above the Pliocene yellow bioturbated sandstone represents the transition from a fan-delta, where the Lucainena flowed into the Plio-Pleistocene sea, to a sequence of alluvial fan deposits. In age, they appear to be the equivalents of the Plio-Pleistocene Gochar Formation of the Sorbas basin. They have been labelled as the Polopos Formation. There are further outcrops of these conglomerates and associated palaeosols exposed in the stream bed and in quarry sections just north of Venta del Pobre.

Unconformably overlying this suite of deposits, and forming what looks like a high river terrace above Polopillos, are further gravels. We interpret

**9.7** Streamside section at Polopillos (Stop 5). This section includes fine yellow sandstones at the base (Pliocene Cuevas Fm.), into which lobate masses of ill-organized matrix-supported conglomerates protrude. We interpret these as the frontal lobes of a Pliocene fan delta. Because of the absence of Filabride hornblende schist clasts, we interpret this as being fed from the Lucainena drainage from the southwest part of the Sorbas basin (i.e., the forerunner of the present drainage) into the corner of the Almeria basin, rather than of the Aguas/Feos system itself. (We saw plenty of evidence of a fan delta fed by that system at the previous site (Stop 4d.) The conglomerate lobes are overlain by bedded conglomerates and then by a palaeosol. We interpret this sequence as representing the transition from a fan-delta setting into a subaerial alluvial fan setting. This sequence is deformed (there is faulting evident in the conglomerates exposed further downstream), then unconformably overlain by sheet conglomerates that culminate in a terrace-like surface. Our colleague Elizabeth Whitfield interprets this as an alluvial fan fed by the Lucainena headwater of the Alias, equivalent to Terrace A further downstream.

these as the equivalent of Terrace A, which here, rather than forming a river terrace confined within a valley, creates a fan-like feature radiating away from the exit of the Lucainena canyon into the northwest portion of the Almeria basin. Quarry sections about 1km SW of this location, which have been studied by Elizabeth Whitfield, corroborate this interpretation. As in many of the river valleys subject to neotectonic activity, the Lucainena valley exhibits some fine incised meandering sections of valley, together with meandering valley cutoffs.

*Return to your vehicle, and to exit this area drive back to Venta del Pobre, where there is a motorway intersection and other roads to Carboneras and to Sorbas.*

### End of Excursion 4

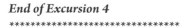

Note: *see* also the extra itinerary linking Lucainena with the Polopos area through the canyon of the Rio Alias across the Sierra de Alhamilla – described in chapter 6 on the Sorbas basin.

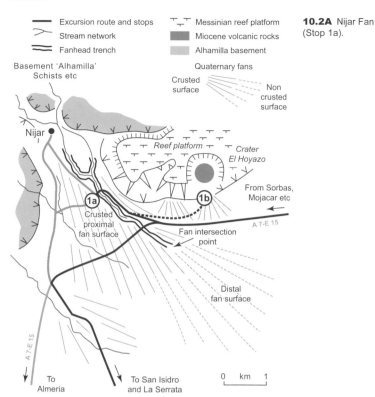

— Excursion route and stops
⤳ Stream network
≡ Fanhead trench

⊤⌐⊤ Messinian reef platform
▓ Miocene volcanic rocks
▒ Alhamilla basement

**10.2A** Nijar Fan
(Stop 1a).

Basement 'Alhamilla'
Schists etc

Quaternary fans

Crusted
surface

Non
crusted
surface

Nijar

Reef platform

Crater
El Hoyazo

From Sorbas,
Mojacar etc

A 7-E 15

1a

1b

Crusted
proximal
fan surface

Fan intersection
point

Distal
fan surface

A 7-E 15

To
Almeria

To San Isidro
and La Serrata

0    km    1

**10.2B** The Nijar alluvial fan looking downfan. The fanhead trench is cut into Azagador Mbr sandstones, overlain by the Quaternary fan conglomerates themselves. The whole is capped by a magnificent calcrete (*see* Fig. 4.6B). The distal area of the fan (Fig. 10.2C) toes out in the centre of the valley (beyond the 'plastic' greenhouses visible in the photo, where it meets small fans fed by La Serrata, the ridge of volcanic rocks aligned along the Carboneras fault system). In the far distance is the main outcrop area of the volcanics, the Sierra de Cabo de Gata.

**10.2C** Nijar fan, distal area, downfan from the intersection point: modern aggradation. Looking downfan towards the terrain described above (Fig 10.2A).

To the northeast the linear reef growth is complicated by what was at the time an eroded volcanic neck (*see* side trip to El Hoyazo that follows this stop) surrounded by further fringing reefs.

The view from the reef platform takes in most of the central and southern part of the Almeria basin. Behind you to the northwest is the Sierra de Alhamilla, here comprising Triassic dolomites of the Alpujarride nappe. The low ridge in the middle distance to the southeast, beyond the 'plastic' agriculture, is La Serrata ridge, formed of a sliver of Cabo de Gata volcanics brought in along the Carboneras faults. In the far distance the hills are the Cabo de Gata range in the volcanics (which we visit later on this excursion). This side of the Carboneras faults the ground comprises coalescent Quaternary alluvial fans, large fans fed from this side from the Sierra de Alhamilla and small fans fed from La Serrata. To the southwest is a relatively uplifted area of the basin comprising offlapping Neogene and Pleistocene sediments, bevelled by Pleistocene pediment surfaces.

*Return to your vehicle and if taking the side trip to El Hoyazo drive back to the motorway intersection, but do not take the motorway. Instead, take the track that parallels the motorway northeastwards, on the north side of the motorway, some 300m before the intersection. Follow this track to a group of abandoned buildings. Park there* [W2.1694, N36.9540] *and walk northeast for about 100m to a small stream bed. Turn left up the stream bed and walk into the 'crater', a depression surrounded by a ring of reef limestone cliffs. This is El Hoyazo, Stop 1b (740907).*

**Stop 10.1b El Hoyazo**

This crater-like landform is an eroded volcanic neck, surrounded by a rim of resistant limestones. The volcanic neck is composed of high-grade calc-alkaline dacites, rich in metamorphic inclusions, the lava having passed up through schistose country rocks. Unusual is the presence of abundant euhedral garnet phenocrysts. These are not only present in the volcanic rock itself, but have been washed out into the small stream draining the 'crater' as alluvial garnets (Fig. 10.3A) , and are also found bedded within the small alluvial fan at the mouth of the 'crater'.

Surrounding the lip of the 'crater' are Messinian fringing reefs. It is easy to envisage a scene, typical of a modern Pacific island, of a coral reef surrounding an eroded volcanic neck. However, due to the relative weakness of the volcanic rock, it has been further eroded to form the modern 'crater-like' landform (Fig. 10.3B), leaving the resistant coral limestone as the surrounding lip.

**10.3** El Hoyazo 'crater' (Stop 10.1b). **A**: Alluvial garnets. **B**: El Hoyazo 'crater'.

*Return to your vehicle. Return to the motorway intersection and take the motorway south towards Almeria for one intersection only. Take the first exit (the southern intersection for Nijar). There, turn left off the motorway and head southeast towards San Jose. You drive down the extensive distal surface of Nijar fan, passing the turn to San Isidro on the left, and into the centre of the basin. From there the road climbs up the surfaces of small fans issuing from La Serrata, then across the first of the Carboneras faults onto volcanic bedrock. The road climbs through broad hairpins to the top. Just before the top, park at the roadside, where there are extensive views north and west. This is Stop 1c: La Serrata (744808)* [W2.1677, N36.8698].

## Stop 10.1c  La Serrata (viewpoint)

The view to the north and northwest from this point encompasses the whole of the western end of the Almeria basin. In the foreground are the coalescent alluvial fans you have driven across en route from Nijar (Fig. 10.4A), small steep fans from La Serrata, much larger fans from the Sierra de Alhamilla, Nijar area, on the north side of the basin. To the northwest is the relatively elevated terrain of offlapping Neogene and Pleistocene sediments, bevelled by early–mid-Pleistocene pediments. Beyond that, to the northwest is the Rioja corridor, between the Sierras Alhamilla (on its eastern side) and Gador (on its western side). The Rioja corridor links through from the distal end of the Tabernas basin (*see* chapter 7, Excursion 2), and the Andarax valley, which in turn drains the terrain between the Sierra de los Filabres and the Sierra Nevada.

**10.4A** The Carboneras faults at La Serrata. Looking north across the Almeria basin towards Nijar and the Sierra de Alhamilla. Coalescent small alluvial fans (in front of you) fed from La Serrata meet much larger fans (fed from the Nijar area) in the zone of 'plastic agriculture'.

**10.4B** Google Earth image of the Carboneras fault system at La Serrata (Stop 10.1c). Movement on the faults is left-lateral. Note how this is reflected in the shifted main streams; however, the picture is not as perfect as textbooks might imply!

Closer at hand, observe the immediate effects of the northern strand of the Carboneras fault system (Fig. 10.4B), that bounds La Serrata to your north, immediately below you. The small streams that drain this side of the ridge are shifted to the left (west), in accordance with the dominant left-lateral displacement along the fault system.

*Return to your vehicle. Continue to drive south over La Serrata ridge, across the second Carboneras fault and back onto Neogene rocks and Quaternary sediments. Turn left at the first roundabout to Los Albaricos. There, turn left towards Fernan Perez (about 10 km from los Alboricos). There, turn right towards Las Negras and drive for about 3 km into the Cabo de Gata range on the volcanic rocks. About 4km beyond Fernan Perez the road climbs through a col, then begins to descend. After a tight left-hand bend, then a right-hand hairpin, park on the shoulder [W2.0317, N36.8787]. Walk a little above the road to where you can clearly see the hillsides to your northeast. This is Stop 2a: opposite Hortichuelas (Cerro Vieuda) – (865824).*

## Cabo de Gata range (the volcanics)

We have described the origins of the volcanic rocks of the Cabo de Gata province in chapter 3, identifying the Rodalquilar caldera as a major structure (*see* Fig. 3.2). On this excursion we first examine the area of the Rodalquilar caldera, then move south into rocks that pre-date the caldera. At the same time we will see the Quaternary features of the coastal zone.

### Stop 10.2a Hortichuelas

The rocks exposed in the hillslope, to your east (Fig. 10.5), form part of the crater wall of the Rodalquilar caldera (*see* Fig. 3.2). The rock in the centre, between the two hills, is a pyroxene andesite lava and breccia. Above this to the left, forming the hill Cerro de la Viuda, are hornblende dacite breccias, and forming the hill to the right (Cerro del Aire) are ignimbrites. Exposed by the road at the base of the sequence are rhyolites and volcaniclastic rocks.

*Return to your vehicle. Continue to drive east towards Las Negras, but at the road junction take the right-hand fork instead, towards Rodalquilar and San Jose. En route you pass the turning for El Playazo, another site with Quaternary raised beach sediments. Rodalquilar lies in the centre of a basin within the Rodalquilar caldera. The rocks within the caldera were affected by intense mineralization, resulting in the emplacement, amongst other minerals, of gold, sufficient to allow the development of a local mine. The village itself relates entirely to the mine. Park on the NW side of the village and walk up to the processing works. This is Stop 10.2b: Rodalquilar (853790: approx)* [W2.0468, N36.8535].

**10.5** Hortichuelas (Stop 10.2a). Dacites, andesites and volcanic agglomerates forming the rim area of the Rodalquilar caldera: younger volcanic series.

## Stop 10.2b Rodalquilar

Walk up to the processing area (Fig. 10.6). The mine is beyond. From the processing area there is an overview of the centre of the caldera. The low hill to the west is a resurgent lava dome formed within the caldera. It is possible to drive through the village to the 'Centro Geoturistico' and 'Punto de Informacion' at Rodalquilar, both of which are located near the start of the mining routes (marked by 'Ruta de la Minería' signs). Both these centres offer information on the flora, fauna, and mining history. Following the tracks it is possible to drive through the mine workings to gain an elevated viewpoint over the mining area. Alternatively, walking is feasible. You are now within the collapse structure (4km by 8km) of the volcanic caldera (*see* chapter 3). The collapse was associated with intense hydrothermal alteration, and the development of gold deposits occurred during the collapse.

The gold is concentrated in 'pipes' of pinkish, low-density aluminium-rich alunite. Sections through some of this local geology can be seen exposed in the track cuttings as you climb to a low col, as well as some old adits (do NOT venture in: some shafts go down to sea level!). The gold was mined initially by deep shafts in the 1930s and became opencast in

**10.6** Rodalquilar gold mine processing plant (Stop 10.2b). The mining area (a pockmarked landscape) is beyond.

the 1940s, ceasing production in the 1950s. In the 1960s some of the previously mined material was reprocessed using improved extraction techniques in the processing area, which still stands today. Approximately 5 metric tonnes of gold were extracted (1940–66). The crushed material and gold was separated using chemical processes involving alkaline (pH=10) cyanide solution and sulphuric acid. The mine tailings from the extraction are still partially visible today. Once they ceased to be maintained, these sites were actively eroding and moving sediment down the gully system towards the coast. There has been some attempt to stabilize and re-vegetate these areas (in 1997) as the sediment included heavy metals (As, Cd, Cu, Pb, Zn) in high concentrations.

*Return to your vehicle. Continue south towards San Jose. After about 2km you emerge above the coast at a viewpoint marked 'Mirador'. This is Stop 10.2c: Amatista viewpoint (856762)* [W2.0394, N36.8278].

**Stop 10.2c  Amatista viewpoint**
This viewpoint gives an excellent view north and south along the coast. To the north the view along the coast is dominated by pyroxene andesite domes of the lower volcanic group. To the northwest La Rellena plateau is formed largely of Cinto ignimbrites. To the south is a large Quaternary alluvial fan (*see* next stop), behind which is the large Frailes volcanic structure, built of a complex of Andesites, at the base the lower volcanic group (12–10Ma), above which they are of the younger volcanic group (8Ma).

*Return to your vehicle. Continue to drive south, past the turn on the left to the hamlet of La Isleta, then across the large Quaternary alluvial fan. Turn left at the next junction to Los Escullos. Park there by the beach. This is Stop 3: Los Escullos (837737)* [W2.0631, N36.8044].

## The coastal alluvial fans of the Cabo de Gata ranges

There are Quaternary alluvial fans issuing from the catchments on the volcanic rocks of the Cabo de Gata range. Those on the landward side are simple fans whose evolution has been controlled entirely by climatically-controlled variations in sediment supply. They exhibit Pleistocene proximal fan surfaces trenched by shallow fanhead trenches which end at midfan intersection points. Beyond the intersection points are untrenched late Pleistocene and Holocene distal fan surfaces (*see* additional stop described at the end of this chapter). Base levels to the inland fans have been more or less constant. On the other hand, the coastal fans exhibit an additional base-level control in response to Quaternary sea-level change. Sea levels were high, of course, as now, during global interglacials, and low during global glacials. Unusually, fan dissection took place *not* at times of low base level (low sea level). At those times the fans were undergoing increased sediment supply and simply prograded onto the then exposed sea floor. During interglacials, sediment supply to the fans was reduced and incision of the fanhead channel took place. These were times of high sea levels when coastal erosion foreshortened and steepened the distal fan channel profiles, accentuating channel incision. Stops 3a and 4b of this excursion visit two such fans.

### Stop 10.3  Los Escullos/La Isleta fan

To the south of Los Escullos are cemented Pleistocene aeolianites, forming fossil dunes (Fig. 10.7A). These preserve root casts on their surfaces. The fossil dunes rest on a thin fossiliferous calcarenite bed (Tortonian–Messinian?) which itself rests on volcanics. Inland to the southwest the dunes are overlain by younger (late Pleistocene or Holocene?) fan and colluvial deposits.

Looking to the north from Los Escullos are low cliffs cut into Quaternary fan deposits. These form part of the large 'La Isleta fan' (Fig. 10.7A,B). We recognize three sets of fan deposits (we have labelled them, oldest

**10.7A** Los Escullos (Stop 10.3). Cemented Pleistocene dunes at Los Escullos, with La Isleta alluvial fan beyond.

**10.7B** La Isleta fan: View downfan from midfan. Note the crusted fan surface (Qf1) and the well-developed soil exposed in section on the left. Los Escullos cemented Quaternary dunes beyond.

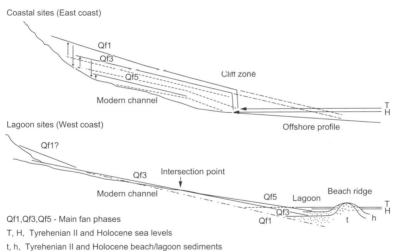

**10.7C** Schematic profiles of coastal and lagoon-based Cabo de Gata fans, illustrating the effects of base-level (Quaternary sea-level) change on the profiles of the coastal sites in contrast with the stable base levels of the lagoon-based fans: successive fan surfaces labelled Qf1, Qf3, Qf5.

to youngest, Qf1, Qf3, Qf5, on the basis of soil development on the fan surfaces and on the basis of their morpho-stratigraphic relationships (Fig. 10.7C,D) The oldest set (Qf1), characterized by a mature soil and calcrete profile, is probably mid-Pleistocene in age and forms the main fan surface.

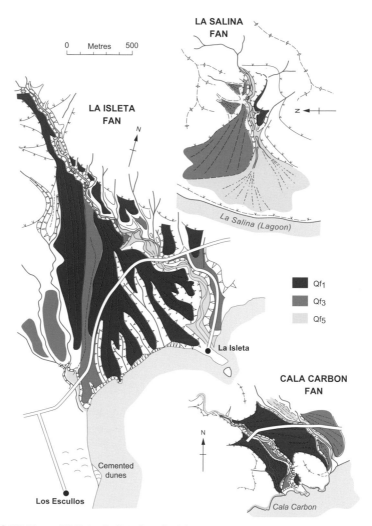

**10.7D** Maps of 3 Cabo de Gata fans (La Isleta, a large east-coast fan – Stop 10.3; Cala Carbon, a small east-coast fan – Stop 10.4b; and La Salina fan, buffered by La Salina lagoon – Stop 10.5a), showing relative age of the fan surface segments (Qf1, Qf3, Qf5).

The intermediate set (Qf3) floors a distributary channel on the southern flank of the fan. The youngest set (Qf5), probably of Holocene age, occurs only in patches on the floor of the modern fan trench. Although at this site we cannot unequivocally relate fan aggradation and dissection phases with the Pleistocene sea-level sequence, the profiles and the morpho-stratigraphic relationships between the bodies of sediment suggest that the fan aggradation phases (Qf1, Qf3) relate to Pleistocene glacials and the dissection phases to interglacials. Dissection has been reinforced by profile foreshortening due to coastal erosion during stages of high sea levels.

*Return to your vehicle. Return to the main (Las Negras–San Jose) road; turn left towards San Jose.*

*En route, after about 1 km, note the large quarry on the right. This is cut into a pyroclastic sequence, involving pumice layers, highly altered by hydrothermal processes into* bentonite. *This quarry and others in the vicinity are the main Spanish sources of cat litter – no doubt a valuable product! Continue towards San Jose, turning left through El Pozo de Los Frailes. On the outskirts of San Jose, rather than descending into the centre of the town, bear right by a petrol station, then right again onto a road signed 'Playas'. This road climbs above San Jose initially as a surfaced (for about 1km) road through suburban villas, then becomes a rough gravel road. Drive slowly, especially in summer when the road can be busy. Drive for about 6km along this road. Note the red soils on the volcanic-sourced Quaternary colluvial deposits on the right. Note also the Quaternary to modern climbing dunes on the hill to the left. Park on the left at the information centre of the Natural Park at Monsul and Media Luna beaches: this is Stop 4a – Playa de Monsul (764656)* [W2.1464, N36.7320].

### Stop 10.4a  Playa de Monsul

The volcanic rocks in the area are thick andesitic breccias, intruded by feeder dykes of andesitic lavas. Walk to the beach. Here the breccias are surfaced by an indurated case-hardened carapace, in which beautiful tafoni are developed (similar to those shown on Fig. 5.1).

*Return to your vehicles and continue southwest along the track for another 1 km to where the track crosses a deep fanhead trench. Park in the small car park. [Incidentially, the track continues along the coast to the Cabo de Gata, but is chained off, and only accessible on foot]. Park here and take the rough track down towards the sea. This is Stop 4b: Cala Carbon (752658)* [W2.1603, N36.7333].

### Stop 10.4b  Cala Carbon

The small Quaternary alluvial fan at Cala Carbon is fed from the west from a catchment in volcanic rocks (Fig. 10.8). The fan itself wraps around a bedrock hill before reaching the coast at Cala Carbon (the present fan outlet) and at another bay a little to the east (Fig. 10.7D). As with other fans in this area, three sets of fan deposits and their associated morphologies

**10.8** Cala Carbon fan: general view looking upfan: Qf1 is the main fan surface, Qf3 the inset surface on the left of the channel. The modern channel is cut down into the underlying volcanic rocks as a result of profile shortening due to coastal erosion. (*See* Fig. 10.7C,D for map and profile.)

can be recognized (Fig. 10.7D). We have labelled them as elsewhere (Qf1, Qf3, Qf5). Virtually the whole surface of Cala Carbon fan is of Qf1 age. As elsewhere it is characterized by a well-developed soil and calcrete. Fan stage Qf3 forms a terrace-like form within the fanhead trench, and exhibits only a weaker soil. Fan stage Qf5 occurs only patchily as sediments on the floor of the modern fanhead trench. The fan is truncated at Cala Carbon by cliffs, which are cut into the underlying volcanic-ash bedrock. Below the cliffs are fragments of cemented beach gravels, presumably, as elsewhere, of MIS 5 age.

The modern profile of the fan trench shows a marked steepening at the seaward end (Fig. 10.7C), suggestive of incision following profile fore-shortening as a result of coastal erosion during high sea levels.

*Return to your vehicles (perhaps after a swim!). Return via San Jose. (You may wish to visit the Cabo de Gata itself; extra itinerary outlined below.)*

## *End of Excursion 5*

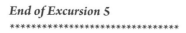

## An extra site in the Cabo de Gata volcanic terrain that is well worth the visit: Cabo de Gata lagoon and headland

*The Cabo de Gata itself is well worth a visit. Approach Cabo de Gata either from the Nijar area or from Almeria. From Nijar follow the motorway south to the Cabo de Gata exit, turning there onto the Cabo de Gata/San Jose road, then west to Cabo de Gata village. From Almeria, leave the city via the airport, then turn right onto the Cabo de Gata road to Cabo de Gata village. In Cabo de Gata village turn left along the coast road towards the small village of Las Salinas.*

The coast road to Las Salinas runs along a shingle structure that separates a large lagoon from the open sea (Fig. 10.9A). The lagoon is used for salt extraction. It is also home to flamingos and other wading birds. According to our colleague, Cari Zazo, sediments on the floor of the lagoon date back to the early-mid Holocene, so the lagoon has been present throughout the present sea-level highstand. Sediments within the barrier beach include not only Holocene sediments but beach sediments dating back to MIS 5. This indicates that the barrier beach/lagoon complex was also present during the MIS 5 sea-level highstand. Therefore throughout the late Pleistocene and the Holocene it is likely to have provided a stable base level to the catchments on the landward side of the lagoon.

*Stop at the south end of the lagoon/barrier beach. Look inland (Stop 5a [W2.2074, N 36.739]).*

**10.9A** Cabo de Gata. La Salina Fan: general view looking downfan. The bar separating the lagoon from the open sea, according to our colleague Cari Zazo, contains beach material deposited during the Last Interglacial marine highstand (OIS5), suggesting that there has been no base-level change affecting the fan since then. Note that the majority of the fan surfaces are of relatively young sediments (Qf3 and Qf5) – *see* Fig. 10.7C,D for map and profile.

### Stop 10.5a  The south end of the Cabo de Gata lagoon

A series of small alluvial fans (Fig. 10.9A), fed from catchments in the volcanic rocks toe out into at the margins of the lagoon. These fans are quite unlike those seen on the east coast (*see* Fig. 10.7C). These fans support only very shallow fanhead trenches that peter out at midfan intersection points (Fig. 10.7D). The proximal fan surfaces are formed on deposits of stage Qf3 (Late Pleistocene); the distal surfaces by deposits of stage Qf5 (Holocene). Any traces of the mid-Pleistocene stage Qf1 are buried by younger deposits.

*Drive south from Las Salinas. The road climbs inland over a steep debris cone, then onto volcanic rocks. Park at the lighthouse at the end of the road (Stop 5b)* [W2.1920, N36.7221].

### Stop 10.5b  Cabo de Gata

From the road to the lighthouse there is a view back along the coast towards Almeria. From the lighthouse there is a spectacular view down onto sea stacks on the modern erosional coast (Fig. 10.9B), comprising andesite lavas. Opposite the lighthouse is a spectacular rock face exhibiting fan-like columnar jointing in the andesite lavas.

*End of extra Itinerary: return to Cabo de Gata village and Almeria or Nijar.*

**10.9B** Cabo de Gata – sea stack in Cabo de Gata andesitic volcanic rocks.

# Chapter 11

# Further afield

While in this area, you may wish to see something of the geology and geo-morphology of the neighbouring areas. Granada would be an appropriate destination for a day trip from the Almeria area. Such a visit should include tourism. The Alhambra palace is a World Heritage Site and well worth a visit. Also the access road to the ski area in the Sierra Nevada reaches into the highest and most spectacular part of the Sierra Nevada, and its Pleistocene (and recent) glacial morphology. There are plenty of general guidebooks available for Granada, so we will not cover such an itinerary here. Instead, we will focus on two day trips to see the geology and geo-morphology of the regions neighbouring Almeria. We suggest one day trip to include the Guadix basin, the eastern end of the Sierra Nevada, and the eastern Alpujarras (all within the Province of Granada), and the other into the Province of Murcia to see the northward continuation of the fault systems, the Guadalentin trough, the Sierra de Carrascoy and the coastal area near Aguilas. Necessarily, the notes we provide are much less detailed than those for the excursions within the Almeria area.

## Excursion 6: The Guadix basin, the Sierra Nevada and the Eastern Alpujarras

### The Guadix basin

The Guadix basin lies to the north of the Sierra Nevada and the Sierra de los Filabres (Fig. 11.1). It is bounded further north by the folded, mostly Mesozoic rocks of the Sub-Betic ranges. Today it is drained by the head-waters of the Rio Fades, a tributary of the Guadiana, through a gap in the Sub-Betic ranges to the northeast of the basin, and into the Guadalquivir depression that lies to the north of the Betic Cordillera as a whole. The Guadix basin fill comprises a stack of mostly Miocene terrestrial sediments culminating in Plio-Pleistocene terrestrial gravels that form extensive sheets in the southern part of the basin. The northern part of the basin is deeply dissected by the Rio Fades.

**11.1** Guadix/Alpujarras route map.

*To leave the Almeria area there are two possibilities. The simplest is to use the A92 motorway through the western end of the Tabernas basin. That road climbs to a col between the Sierra de los Filabres and the eastern end of the Sierra Nevada near Gergal. The alternative is the old road that follows the Andarax valley through Gador, meeting the motorway near Gergal. Continue west from Gergal through the pass near Dona Maria either on the A92 motorway or on the old road through Ocana and Alba to Finana on the margins of the Guadix basin. If on the old road, rejoin the A92 motorway just before Finana, and continue on the motorway towards Guadix.*

It is interesting that the drainage from this corner of the Guadix basin escapes to the southeast through the Dona Maria pass into the Naciemiento river and the Anadarax ultimately into the Almeria basin. It appears to have been initiated as a flanking drainage along the margins of a Plio-Pleistocene alluvial fan fed southwards from the Sierra de Baza, the far western end of the Sierra de los Filabres. The drainage is orientated towards the southeast and the gap between the Guadix basin and the Andarax drainage system, rather than towards the north and the basin centre.

Beyond Finana, the Sierra Nevada are to your left (south) and the undissected surfaces of the Guadix basin open out to your right (north). The surface is formed of the top of the Neogene basin fill (Fig. 11.2A) overlain by Plio-Pleistocene conglomerates deposited as large fans. Surface soils are deep red. Later you will have a chance to see these sediments. As you approach Guadix you begin to descend into the deeply dissected centre of the basin, dissected and drained by the Rio Fades.

*As you approach Guadix you have two alternative routes. Either leave the motorway by the N324 through Alcudia de Guadix and towards Guadix city centre. At the large roundabout approaching the centre of Guadix take the right-hand turn, the N342 towards Baza (alternatively labelled the A92N). As an alternative route, continue on the A92 to the intersection with the A92N, turning right here towards Baza. After about 15km, exit onto the GR6100 towards Gorafe. Continue for about 6km towards Gorafe to where the road begins to descend into the valley. Park on the shoulder, just before the road begins to descend* [W3.0196, N37.4489].

**11.2A** The Guadix Basin. SE of the basin, near Calahorra. Looking north from the lower slopes of the eastern Sierra Nevada across the undissected southern part of the Guadix basin: note, you will pass this locality later during the excursion, en route across the Sierra Nevada.

The road to Gorafe is initially on the plateau surface on the Plio-Pleistocene conglomerates (studied by our colleague Cesar Viseras), but then begins to descend into the Gor valley. From the lip of the descent (Stop 11.1a) it is possible to examine the Plio-Pleistocene conglomerates capping the basin-fill sequence. The view across the valley takes in the far canyon walls, which expose much of the Miocene sequence of terrestrial basin-fill deposits (Fig. 11.2B). The valley itself is deeply dissected below the plateau surface, and feeds towards the northern part of the basin, deeply dissected by the Rio Fades.

**11.2B** Dissection below the top surface of the Guadix basin: the Gor valley above Gorafe. The basin fill is dominantly of Miocene continental sediments, capped by Plio-Pleistocene fluvial and alluvial fan sediments, surfaced by a mature soil and well-developed calcrete.

**11.3** The Fades Valley. **A**: Tufa formations: Balneario de Alicun de las Torres. **B**: Hillslope badlands resulting from the rapid incision of the Fades Valley into the Miocene basin-fill sediments. Fades valley upstream of Alicun de las Torres.

Beyond Gorafe the road descends into the Fades valley. On the edge of the valley is a hotel/resort (Balneario de Alicun de las Torres) built around a set of hot springs. There are impressive tufa formations, including elevated tufa-lined channels (Stop 11.1b)(Fig. 11.3A) [W3.1077, N37.5088].

*About 1 km beyond the hot springs turn left (up the Fades valley) onto the GR 5103, then later, left again onto the A325 through Fonolas to Benalua.*

The deeply dissected valley sides of the Fades expose much of the basin-fill sequence, including the Tortonian marine marly sediments in the lower part of the fill, then Messinian to Pliocene terrestrial silts, sands, and conglomerates. Note the badland development, with a range of badland styles (Fig.11.3B) on the different materials.

*Past Benalua rejoin the A92 motoway, heading back east towards Guadix and Almeria. You bypass Guadix and continue on the motorway for about 17 km to the exit for the A337 for La Calahora. Turn south onto this road towards La Calahora.*

## The eastern end of the Sierra Nevada

The Sierra Nevada are formed of metamorphic rocks of the Nevado-Filabride nappe, thrust forward northwards towards the Guadix basin, and further west, towards the Granada basin. The highest elevations are at the western end, south of Granada. The Sierra Nevada is bounded to the south by the structural depression of the Alpujarras, separating the Sierra Nevada from the Sierras de Contraviesa in the west and the Sierra de Gador in the east, both ranges formed predominantly of the Alpujarride nappe. The Alpujarras remain a deeply dissected, structurally-controlled valley system that never really developed into a sedimentary basin.

*Continue south on the A337 past La Calahora, where the road swings east, then takes a right turn south into the Sierras towards Laroles, Picera and Cherin.*

As you climb above La Calahora (note the fortress), note also the spectacular views back over the undissected southern and eastern part of the Guadix basin (Stop 11.2a) (Fig. 11.2A). The road then climbs through a series of hairpins through the forested lower slopes of the Sierra Nevada. There are tantalizing views to the west giving occasional glimpses of the highest parts of the Sierra Nevada (Stop 11.2b) (Fig. 11.4), but eventually these are lost behind the mass of the mountain immediately to your right (west). The summit of the road is rather disappointing [W3.0339,

**11.4** Tantalizing partial views west towards the higher parts of the Sierra Nevada, from the road south of Calahora.

N37.1208], in a pass between mountains. South of the pass the views open out (Stop 11.2c). The road runs along a high spur before descending steeply down a series of hairpins to Laroles, a charming white village perched above the central Alpujarra valley. The central Alpujarras are drained by the deeply incised headwaters of the Adra system that flows due south between the Sierra de Gador (to the east) and the Sierra Contraviesa (to the west), to the coast at Adra.

*Continue down more hairpins to Picera and Cherin, where you turn left (east) onto the A603 towards Alcolea and the eastern Alpujarras, then turn onto the A348 to Fondon.*

### The Eastern Alpujarras

The Alpujarras is a structural depression between the Sierra Nevada to the north, and to the south the Sierras De Gador (in the east) and the Sierra Contraviesa (in the west). You are descending steeply into the Adra drainage, which runs towards the south, exiting the Alpujarras between the Sierras Contraviesa and Gador. After Alcolea the road climbs out of the Adra drainage into the Andarax (west-to-east) drainage. In the divide area is a set of little-dissected Quaternary alluvial fans (Stop 11.3 [W2.9264, N 36.9817]) (Fig. 11.5A) (studied by our colleague Tony Garcia), but to their east beyond Fondon, the drainage towards the Andarax becomes increasingly incised (Fig. 11.5B), and sediment accumulation is very limited.

*Continue east along the A348 east through Almocita, Canjaya and Rogol, eventually to Alhama de Almeria.*

**11.5** The eastern Alpujarras. **A**: The little-dissected zone in the upper Andarax valley (eastern Alpujarras) comprising coalescent Quaternary alluvial fans. **B**: The lower Andarax canyon, cut into Pliocene silts.

The entrenched channel of the Rio Andarax, with its incised meanders, lies well to the south of the road until you pass Canjaya. There, you cross the river and pass from the basement schists onto the Pliocene Gador conglomerates of the lower Andarax valley (*see* chapter 7, Excursion 2). Beyond here the metamorphics of the Sierra de Gador are on your right (south), and the dissected terrain of the lower Tabernas basin is on your left (north and east).

Note that there is a parallel with the Tabernas basin: little Quaternary dissection in the upper part of the basin, but deep, tectonically induced dissection in the lower part of the basin.

*From Alhama de Almeria continue to Benahadux. There, either turn left to Rioja and the A92 motorway to Tabernas (for Sorbas), or continue south to Almeria.*

## End of excursion 6

*******************************

# Excursion 7: Southern Murcia: basin and range terrain

The strike-slip fault system that dominates the structure of Almeria continues north (as part of the Trans-Alboran shear zone) into the province of Murcia (Fig. 11.6). The terrain is dominated by a series of NE–SW linear depressions and basins, bounded and separated by mountain ranges. The mountain ranges are formed largely of metamorphic rocks of the Alpujarride nappe, though there are some ranges composed of Nevado-Filabride material. The basin sediments range from the upper Miocene to the Plio-Pleistocene. During the Quaternary, alluvial fans have been

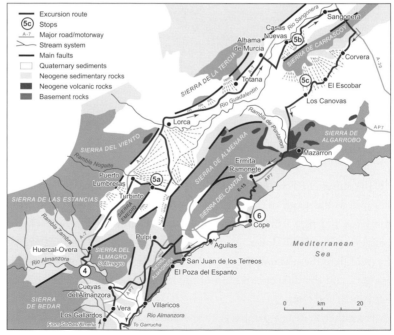

**11.6** Southern Murcia/Guadalentin corridor, Sierra de Carrascoy. Aguilas, Pulpi corridor map.

deposited at the mountain fronts. In some places aggradation has continued in the basin centres, elsewhere the drainage network has incised through the Neogene basin-fill sediments.

*From the Almeria area take the A7 motorway north towards Murcia. Continue along the A7 motorway across the northern side of the Vera basin to the crossing of the Almanzora at El Pilar/Santa Barbara [W1.9673, N37.3470].*

### Rio Almanzora at Santa Barbara (Stop 11.4)

It is worth leaving the motorway at El Pilar/Santa Barbara and to drive back south ~500m on the old road to the bridge over the Rio Almanzora (916340). Upstream from this point the Almanzora that has drained the large Huercal Overa basin has a wide alluvial channel, but at this point the channel begins to be bedrock-constrained as the river enters the canyon section through the Sierra de Almagro (Fig. 11.7). *See* chapter 8, Excursion 3 for an account of the transverse course of the Almanzora.

*Either rejoin the A7 motorway and head north towards Puerto Lumbreras and Murcia or stick to the old road (the N340a) through Huercal Overa, rejoining the A7 motorway north of Huercal Overa en route for Puerto Lumbreras.*

**11.7** Looking into the start of the Almanzora canyon, from Santa Barbara. (Note that the former main road bridge was destroyed by a large regional flood event in 1973.)

### The Guadalentin trough to the Sierra de Carrascoy

One advantage of taking the old road is that it allows you to see the deep dissection of the drainage, tributary to the Almanzora, the Rambla Zambra, cutting into the basin-fill sediments of the southern end of the

Guadalentin trough. As you leave Huercal Overa you climb up onto the undissected surface of the Guadalentin trough, fault-bounded to the north by the eastern part of the Sierra de las Estancias and to the south by the Sierra Enmedio. The faults are part of the series of the NNE–SSW trending Trans-Alboran Shear zone, and define the major outlines of the terrain of this part of Murcia and Almeria provinces. The result is a series of parallel trough-like basins separated one from another by fault-bounded mountain ranges. The southern part of the Guadalentin trough, north almost as far as Lorca, is an area effectively of modern internal drainage, fed by coalescent mountain-front alluvial fans (Fig. 11.8). The A7 motorway, south of Puerto Lumbreras, crosses a series of alluvial fans in mid-fan. The largest of these fans is fed by the Rambla Nogalte, which enters the trough at Puerto Lumbreras. To the east of Puerto Lumbreras is a large, low-angle fluvial fan on which the modern drainage peters out in a series of distributary channels within an area of modern aggradation.

*To see something of these features, leave the A7 motorway at the first exit for Puerto Lumbreras, south of the town. Head into town on the old N340, but as you enter the built-up area turn right onto the RM-D19 towards Tebar and Aguilas. This road runs through low hills at the northern end of the Sierra Enmedio, composed of Nevado-Filabride schists, then skirts the southern margin of the Nogalte terminal fan. If you wanted to actually see the distributary channel network, it is most easily accessed beyond the railway bridge at*

**11.8** The southern end of the Guadalentin trough, looking SE from south of Puerto Lumbreras. The hills in the middle distance are the low northern end of the Nevado-Filabride Sierra Enmedio. The skyline is the Nevado-Filabride Sierra de la Almenara. The trough is filled with coalescent Quaternary alluvial fans.

*Turbinto (104547). Just over a kilometre beyond Turbinto, turn left onto a track heading north for about a kilometre into the zone of the distributary channels of the Nogalte fan (Stop 11.5a [W1.7323, N 37.5327]). Return to the Puerto-Lumbreros to Aguilas road; turn left for another 2.5 km or so (143536). Turn left again onto the RM620 to La Escucha (162575) where you turn left again towards Lorca. This road takes you through the basin centre across the far distal portions of fans derived primarily from the west. On approaching Lorca it is best to make your way back to the A7 motorway, which bypasses Lorca, a very congested town. Continue on the A7 motorway north towards Murcia, past Totana, to exit at Alhama de Murcia onto the RM-2 towards Cartagena.*

The Rio Guadalentin enters the trough at Lorca, and unlike the Nogalte is a through drainage linking north to the Segura at Murcia. According to our colleague Pablo Silva, the Guadalentin used to flow through the gap between the Sierras de la Almenara and de Carrascoy into the Puntaron basin towards Mazarron in the western part of the Campo de Cartegena, but during the late Pleistocene was captured by the Sangonera, working headwards along the axis of the Guadalentin trough from the Segura. A further human-induced partial diversion took place during the nineteenth century AD, when a flood channel was cut along the former drainage line. This is effective only at times of high flows.

*Follow the RM-2 east for one exit (433850). Exit onto the RM-603, turning left and heading north towards Sangonera, along the foot of the large steep alluvial fans issuing from the Sierra de Carrascoy to the toe of Roy Fan: Stop 11.5b (503945), [W1.2897, N37.8856].*

The Sierra de Carrascoy has a core of Alpujarride metamorphic rocks, flanked towards its eastern end on both northern and southern sides by Cenozoic rocks. At the western end of the mountains, both northern and southern margins of the mountains are flanked by Quaternary alluvial fans. Those on the north are spectacular, issuing from steep catchments on the thrust-faulted north face of the Sierra de Carrascoy (Fig. 11.9A,B,C). Exposed in the fanhead trenches are faults, affecting mid- and late-Pleistocene fan deposits. Direct access is now almost impossible, as most of the fan surfaces are now devoted to citrus cultivation. Our colleague Pablo Silva has studied the tectonic geomorphology of the Guadalentin trough, and has identified that in addition to fan-feeder catchment size, one of the main factors affecting the geometry of the alluvial fans is the tectonic setting of the mountain front. The Carrascoy fans are on a tectonically active thrust

**11.9** Sierra de Carrascoy (N): alluvial fans at the faulted mountain front of the Alpujarride range. (Photos taken before the extension of citrus cultivation over the fan surfaces.) **A**: Roy Fan, looking upfan: Carrascoy northern mountain front. **B**: Roy fan: fanhead trench. Note the section in calcrete-crusted Quaternary fan sediments. These are affected by mountain-front faulting. **C**: Roy fan, viewed from above (from slopes of the Sierra de Carrascoy). Note the fanhead trench (centre right) and the intersection point (centre left).

mountain front. They are fed by steep mountain catchments. The fans are simple steep mountain-front fans. Even if difficult of access now, it is worth pausing on the road at Stop 11.5b between Casas Nuevos and El Canarico to appreciate the tectonic situation.

*Continue along RM-603 past Sangonera la Verde to join the MU-31 south-east to the A30 motorway across the Sierra de Carrascoy. Continue along the A30 (direction: Cartegena) to the next intersection onto the RM-601 heading southwest towards Corvera. From Corvera turn right heading southwest on the RM-E6 to Los Paganes between El Escobar (570820) and Los Almagros (530810). This is Stop 11.5c [W1.2492, N37.7667].*

## Campo de Cartegena, the Cope basin and back to Almeria through the coastal region

The Sierra de Carrascoy is formed of the Alpujarride nappe. On the south side it is overlain by a sequence of Cenozoic rocks, culminating in calcrete-capped Pliocene silts that form a low but distinct escarpment. Between the back side of the Carrascoy and the escarpment is a depression occupied by Quaternary fans issuing from the south side of the Sierra de Carrascoy. The RM-601 road to Corvera and beyond climbs over the escarpment, then descends onto the toe of the fan complex emanating from the southern side of the Sierra de Carrascoy. The fans are best seen to the right (north) of the road between at Los Paganes between El Escobar and Almagros. The fans are trenched throughout most of their length (Fig. 11.10A), often inter-secting buried calcrete layers (Fig. 11.10B; either former fan surfaces or groundwater calcretes?). These often form distinct headcuts in the channel floors (Fig.11.10A). The fan surfaces support exceptionally well-developed pedogenic calcrete crusts.

*From Los Pagenes continue through Los Almagros and Canovas. Continue heading WSW for about another 7km to the RM-3 Totana–Mazarron motor-way (452749).*

In this area the caprock escarpment, formed by a calcrete over Pliocene red silts, is impressive.

*Turn south onto the motorway towards Mazarron. After crossing the AP7 motorway there are two possibilities. Either join the AP7 (a toll motorway) and turn south towards Aguilas. Alternatively, take the 'scenic route'. Continue southwards to the next intersection and there turn right (west) onto a minor road (the RM-332) which runs west along small fans at the foot of the Sierra de*

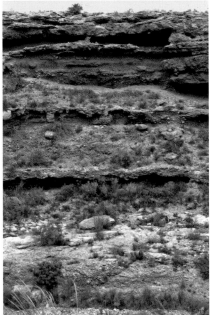

**11.10** South of the Sierra de Carrascoy. **A**: Carrascoy southern fans, near Los Paganes: midfan headcut, exposing heavily calcreted Quaternary fan sediments. **B**: Quaternary fan sediments exhibiting multiple calcretes (section height *c*.5 m).

*las Moreras. The road then turns south across a col (in a tunnel, parallel with the AP7). Continue west along the RM-332 to Ermita Ramonete. If on the AP7 rather than the RM-332, there is an exit for Ermita Ramonete. In both cases, turn left after the village onto the RM-D20 over the mountains (spectacular views from the col at 360525) (Fig. 11.11A) into the head of the Cope basin* [W1.4581, N37.5110]. *For access to the sea (Fig. 11.11B) and to sections in*

**11.11** Cope basin. **A**: The Cope basin, viewed south from the pass over (Nevado-Filabride) Sierra del Cantar. The headland bounding the basin to the south is of a thrust-in Jurassic limestones. **B**: The Cope coast: low cliffs cut Pliocene shallow-marine sediments, overlain by (very thin here) Quaternary calcreted gravels.

*the Neogene basin-fill sediments, as well as to Pleistocene raised beach deposits, about 1km short of Los Alcazar turn left onto the RM-D15 to the coast. This road follows the coast, then from Torre de Cope it will take you through Calabardina towards the RM-D14 and Aguilas.*

West of Mazarron are low mountain fronts, with a variety of geology, including Neogene volcanic rocks, similar to those in the Vera basin (*see* chapter 8). The low mountains are fronted by calcrete-crusted fans. Now out of the Mazarron area, the Cope basin (Stop 11.6) is a small sedimentary basin of Neogene basin-fill sediments overlain by Quaternary alluvial fans, open to the sea on the east. Cope headland is a block of Liassic Limestone, thrust from the south.

*If pushed for time and wishing to skip the visit to the Cope basin, either follow the RM-D20 and RM-D14 directly from Ermita Ramonete to Aguilas, clipping the corner of the Cope basin, or missing the Aguilas area altogether, stay on the AP7 towards Pulpi.*

Aguilas is at the centre of the Aguilas arc (Fig. 11.12), a compressional structure, created by northward movement along the Palomares fault system comprising rocks of the Alpujarride nappe in the centre (Sierra del Cantar) thrust against Nevado-Filabride rocks in the surrounding Sierra de la Cuiscuilla, along the northern part of the Palomares fault system.

*From Aguilas there are three alternative routes back to the Almeria area. In each case take the Aguilas bypass (the RM-333) around Aguilas.*

i) *Take the RM-11 from the NW corner of the Aguilas bypass, back to the AP7 Expressway, then follow the AP7 expressway (toll) south. This will take you through the Pulpi basin, across the Almanzora to join the A7 motorway near Vera.*

ii) *Continue to the west end of the Aguilas bypass on the RM-333 to Calarreona then along the coast road, the N332. At the provincial border take the road bypassing San Juan de los Terreros to turn north on the Al-350 across the*

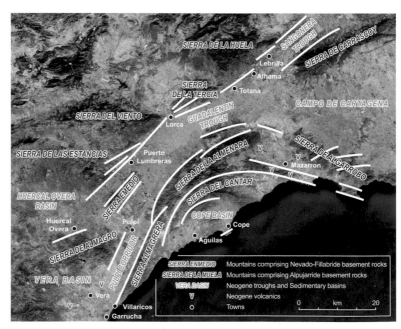

**11.12** Google Earth image of the Aguilas arc, showing the main structural units.

*northern end of the Sierra Almagrera into the Pulpi basin. In Pulpi turn south onto the AL-610, then through that basin, following the Palomares fault, across the Almanzora to Cuevas del Almanzora and Vera in the Vera basin.*

iii) *Follow route (ii) above as far as San Juan de los Terreros, then take the coast road (AL-1065) along the edge of the Sierra Almagrera to Villaricos, across the Almanzora to Garrucha in the Vera basin.*

The Sierra Almagrera is a mountain group of dark Alpujarra schists, the detached northward continuation of the Sierra Cabrera. It was tectonically transported north along the Palomares faults, to be thrust into the Nevado-Filabride Sierra de la Cuiscuilla to its north. The Palomares fault system lies to the west of the range, bounding the Pulpi basin. The basin fill of the Pulpi basin is dominated by Pliocene shallow-marine sediments. The basin is flanked to the east by Quaternary alluvial fans emanating from the Sierra Almagrera.

Note again that there is a parallel with the Almeria region. Most of the basin centres in the Murcia 'basin and range' terrain, as zones of tectonic sag, are non-dissecting, aggrading basins. Only around the basin margins is there limited dissection in the form of fanhead trenches, tectonically controlled, but climatically modified. Only near a deeply incising major river, or adjacent to the coast, does the local base-level effect override the overwhelming tectonic signal.

***End of Excursion 7***
*********************************

# Appendix 1

# The Geological Timescale

| | | | AGE (Million yrs BP) |
|---|---|---|---|
| Quaternary | | Holocene Pleistocene | 1.8 |
| Cenozoic (Tertiary) | Neogene | Pliocene Miocene | 23 |
| | Palaeogene | Oligocene Eocene | 65 |
| Mesozoic | | Cretaceous Jurassic Triassic | 250 |
| Palaeozoic (Upper) | | Permian Carboniferous Devonian | 415 |
| (Lower) | | Silurian Ordovician Cambrian | 540 |
| Precambrian (Proterozoic) (Archaean) | | | |

# Appendix 2

# Neogene to Quaternary Stratigraphy of the Sedimentary basins of Almeria

| Era | Epoch | Tabnernas | Sorbas | Vera | Almeria |
|---|---|---|---|---|---|
| Quaternary | Holocene | Floodplains, fans | Floodplains, E terrace | Floodplains | floodplains, fans |
| Quaternary | Pleistocene | fans, high terraces | A,B,C,D terraces | terraces | fans, terraces |
| U. Miocene / Neogene | Pliocene | | Gochar Fm. | Salmeron Fm. | Polopos Fm. |
| U. Miocene / Neogene | Pliocene | | Carietiz Fm. (Zorreras/Moras Mbrs.) | Espiritu Santo Fm. Cuevas Fm. | (Espiritu Santo Fm.)? Cuevas Fm. |
| U. Miocene / Neogene | Messinian | (Yesares) (Cantera Mbr.) Azagabor Mbr. | Turre Fm Yesares Mbr. Cantera Mbr. – Abad Marl Azagabor Mbr. | Turre Fm Azagabor Mbr. | (Yesares Mbr.) Cantera Mbr. Azagabor Mbr. |
| U. Miocene / Neogene | Tortonian | marls/tubidites | marls/tubidites | marls/tubidites | marls/tubidites |
| U. Miocene / Neogene | Serravallian | conglomerates | conglomerates | | |

Basin ⟶

# Appendix 3

# Quaternary timescales, related to the Neogene sedimentary basins of Almeria

| Years BP (Approx) | Global context | MIS stage (OIS) | Previous terminology | Significant events in Almeria — Coast | Terrace and fan sequences | Other |
|---|---|---|---|---|---|---|
| 10-0 ka BP | Interglacial | 1 | Holocene | Modern sea level (since c 6ka BP) | Incision, Floodplains Youngest terraces (E) Fan trenching, youngest fan surfaces | |
| 30-10 ka BP | Last Glaciation | 2 | Wurm | Low sea levels | Terrace D3 Fan deposition | Drainage of the Tabernas "lake" - Rapid incision |
| 60-30 ka BP | Interstaial | 3 | Mid-Wurm | Sea levels below present | incision | |
| 80-60 ka BP | Glacial | 4 | Early Wurm | Low sea level | Terrace C Fan deposition | Possible initiation of Tabernas lake? Providing base level for last of the "badland" pediments |
| 130-80 ka BP | Last Interglacial | 5 | Tyrhenian II Marine stage | High sea levels Main raised beaches | Incision Fan trenching | |
| 200-130 ka BP | Glacial | 6 | Riss? | Low sea level | Terrace B Fan deposition | Tabernas "badland" pediments? |
| 250-200 ka BP | Interglacial | 7 | Tyrhennian I ? Marine stage | High sea level Higher raised beaches | Incision? Fan trenching | |
| 300-250 ka BP | Glacial | 8 | ? | Low sea level | Fan Deposition | |
| >300 | MID PLEISTOCENE Glacial/Interglacial sequences | | Earlier marine stages | Oscillating sea levels | Oscillating Aggradation/dissection sequences Terrace A? | |
| 1.8Ma BP | EARLY PLEISTOCENE | | | | Gochar surfaces Gochar formation:deposition | Tabernas high level terraces |

# Appendix 4

# Glossary

## A

**Accommodation space**: available vertical and/or lateral space for the deposition and accumulation of sediment.

**Aeolian**: by wind. Aeolian deposition: deposition of windblown sediment.

**Aeolianite**: cemented dune sand, usually cemented by $CaCO_3$, and occurring mostly in warm dry regions, often fossilizing former coastal dunes.

**Agglomerate**: a sedimentary rock composed of relatively angular rock fragments (compare with: *conglomerate*).

**Aggradation**: the (usually) vertical accumulation of sediment.

**Alluvial fan**: a conical or sub-conical depositional landform, deposited where a steep stream leaves the confinement of a mountain catchment, either at a mountain-front or a tributary-junction setting. Common in arid mountain areas, but occurs under virtually all climatic conditions. Includes a range of sizes from tens of metres to tens of kilometres in length. Sedimentation processes range from debris flows to fluvial processes, either as sheetflows or as streamflows.

**Alluvial channel**: a river channel whose margins comprise alluvium (i.e. previously deposited river sediment). Processes in such channels usually involve a balance between erosion and deposition.

**Alluvium**: sediment deposited by a stream or a river. Typically implies fine sediments, eg. silts, but strictly includes all grain sizes.

**Anabranch**: a stream or river channel within a channel network that divides and rejoins (see: *Anastomosing*, and *Braided channels*).

**Anastomosing river channel**: a low-energy divided river channel, usually in fine (silty/muddy) sediment; (compare: *Braided channel*).

**Andesite**: a fine-grained crystalline intermediate igneous rock occurring primarily as lava and as minor intrusions, composed primarily of plagioclase feldspar and hornblende.

**Angle of rest**: depositional slope angle at which clasts of a given size come to rest.

**Anhydrite**: the anhydrous form of gypsum $(CaSO_4)$.

**Antecedent drainage**: drainage lines that cross active fold structures, most obviously anticlinal folds, such that the rate of fluvial incision keeps pace with the rate of uplift, resulting in a drainage pattern transverse to structure.

**Anticline**: a fold in bedded rocks in the form of an arch. The upper (younger) rocks in the centre may be removed by erosion, exposing the older rocks underneath in the form of a breached anticline.

**Augen gneiss**: a medium- to coarse-grained, banded high-grade metamorphic rock (see *Gneiss*), composed of quartz, feldspar, hornblende and mica, in which there are eye-shaped clusters of feldspar crystals aligned with the banding in the rock.

**Augite:** A ferro-magnesian mineral characteristic of intermediate-to-basic igneous rocks.

**Avulsion:** river channel change by diversion and spillage during flood conditions into a backswamp environment adjacent to a river channel or onto a floodplain or alluvial fan surface, often causing the original channel to be abandoned. The new channel may rejoin the original channel further downstream, or, as is especially the case on alluvial fans, may represent a completely new direction of drainage.

## B

**Badlands**: intensely gullied hillslopes, especially common in drylands on weak, easily erodible marl or shale bedrock, usually with little or no vegetation cover. They are usually characterized by a very high drainage density of rills and gullies. In some areas badland development is the result of intense human-induced soil erosion.

**Basalt**: a common volcanic rock, dark in colour, usually formed as lava flows. A basic as opposed to an acid composition, dominated by ferromagnesian minerals (pyroxenes and olivine), together with some feldspar. May be characterized by columnar hexagonal joints, contraction cooling cracks.

**Base level**: the lower elevational limit of subaerial erosion processes. May be local, as in the case of a resistant rock horizon or a main valley floor, or may be regional, as in the case of sea level. A base-level fall is a major cause of drainage incision.

**Bedrock channel**: a river channel cut into bedrock, in which stream power is too high to allow the accumulation of sediment (in contrast with *alluvial channels*).

**Bentonite**: an aluminium silicate mineral that results from the decomposition of volcanic ash, predominantly clay-sized particles and highly absorbent.

**Bioturbation**: the disturbance of primary sedimentary structures within a sediment or sedimentary rock, by animal or plant activity.

**Blind fault**: a buried fault: usually in basement underneath cover rocks. Movement on the fault deforms the cover rocks.

**Bottomset**: near-horizontal beds accumulating in front of an accretionary sediment wedge.

**Braided river channel**: a multi-thread river channel in which the channel divides around sand bars, gravel bars or vegetated islands. Characteristic of high-energy bedload dominated rivers. Distinct in form and behaviour from lower energy, lower gradient, mud-dominated, *anastomosing channels*, which are much more stable.

**Breccia**: a sedimentary rock composed of angular stone-sized particles: *Brecciation* refers to the *in situ* fracturing of a massive rock into angular fragments by weathering processes. A common phenomenon in mature *calcretes*.

## C

**Calcarenite**: calcereous sandstone.

**Calcite**: a mineral, $(CaCO_3)$, soluble in weak acids (rainwater, humic acids), the primary constituent of limestone, responsible for the dominance of solutional forms in limestone (karst) regions.

**Calcrete**: an *Indurated* layer formed of calcite. Pedogenic calcrete forms by the downward movement of $CaCO_3$ and its precipitation some way below the surface as a Bk or K horizon in areas with a sustained soil moisture deficit, i.e. semi-arid regions. On exposure, such horizons become indurated, undergoing a complex sequence of brecciation and recementation. They may form a duricrust or caprock. Groundwater calcrete forms in similar areas, but in zones of the soil profile just above the water table or where there is a reduction in vertical permeability. *See* also *Duricrust*.

**Caldera**: a large volcanic, crater-like form produced by collapse.

**Cantilever structure**: laterally supported resistant rocks, creating an overhang. On collapse, results in a tilted slab of rock.

**Canyon**: incised river valley or gorge, characteristic of rapid incision in dryland regions.

**Caprock**: a resistant layer that may be a resistant rock layer or a *duricrust,* protecting a hilltop or escarpment from erosion, underlain by a weaker layer more prone to erosion, the two giving a cliff face over a concave slope morphology.

**Capture**: see *River capture.*

**Case hardening**: the concentration of minerals in the outer layer of a rock by their precipitation through desiccation. Protects the outer surface of the rock from erosion. May lead to the formation of *Tafoni* (honeycomb weathering).

**Chemical weathering:** the chemical breakdown of rock-forming minerals (in igneous, metamorphic or sedimentary rocks) by chemical processes (particularly by solution, oxidation/reduction, hydration, hydrolosis) to form (within the particular chemical environment) chemically more stable compounds.

**Chlorite**: a hydrous aluminium silicate mineral, usually green in colour. It occurs in igneous rocks by the alteration of ferromagnesian minerals. It is also common in metamorphic rocks.

**Chronosequence**: a set of related soils whose main differences are due to age differences: useful in estimating the relative ages of geomorphic surfaces.

**Clasts**: rock fragments.

**Clinoforms**: inclined or dipping features (beds) within a sedimentary rock sequence.

**Colluvium**: ill-sorted fine sediment accumulating at the foot of hillslopes through diffuse soil erosion or creep processes.

**Conglomerate**: a sedimentary rock composed of rounded pebble- to cobble-sized clasts.

**Consequent drainage**: initial drainage pattern, created on a newly exposed or tectonically uplifted surface, in which the drainage direction follows the original slope.

**Continental crust**: the thicker crust under continental areas (including continental shelves), composed of less dense granitic rock, as opposed to *oceanic crust,* which is composed of basaltic rock.

**Coupling**: linkages along a river channel or between slopes and channels, such that sediment may move down through the system, or incisional waves can propagate up the system.

**Cross bedding**: within a sedimentary bed, laminae or bedding surfaces deposited at a sloping angle to the main bedding units.

**Cutoff**: an abandoned meander.

## D

**Dacite**: a fine-grained, crystalline, mildly acidic to intermediate igneous rock occurring primarily as lava and as minor intrusions, composed primarily of orthoclase feldspar and biotite and hornblende.

**Debris flow**: a mobile mass of unconsolidated rocks and fine sediments with variable water content moving downslope, often within a gully. Its source may be a slope failure, or may simply be within-gully sediment. Movement is usually triggered by heavy rains. Movement ceases as slopes decrease. On deposition may form a debris-flow lobe; important constituents of alluvial fans from small, steep catchments. Can be very hazardous to human life and settlements, especially in the case of *lahars*, debris flows triggered by volcanic activity.

**Dendritic drainage**: a randomly branching drainage network.

**Diapiric**: describes a process whereby light material buried at depth rises through the surrounding material. The process of salt-dome formation involving halite $(NaCl)$ is well known. There is some debate as to whether gypsum $(CaSO_4. nH_2O)$ behaves diapirically – it may require a tectonic trigger.

**Dolomite**: magnesian limestone.

**Duricrust**: an *indurated*, erosionally resistant layer at or near the surface, overlying weaker sediment or soil, often forming a 'caprock'. Formed by chemical precipitation of calcium carbonate (to form calcrete), iron oxides (to form ferricrete) or silica (to form silcrete).

## E

**Epeirogenic (post-orogenic) uplift**: regional tectonic uplift of a zone of crustal thickening produced by past plate-tectonic activity in a mountain belt (e.g., after subduction ceased in the European Alps in the Miocene the whole region has been uplifted during the Pliocene and Pleistocene). Additionally, upwelling mantle processes may cause regional uplift (e.g. as in the Colorado Plateau, USA).

**Erosion surface**: this term has two meanings – in a sedimentological sense: a relatively small-scale feature, an erosional horizon cross-cutting the sedimentary structures of an underlying deposit. At a larger scale: extensive surfaces that cross-cut the underlying geological structure, often forming upland plateaux, characteristic of the 'older' mountain belts, the 'uplands' of Western Europe. There is controversy concerning their origin, whether they are uplifted former *peneplains, pediplains,* marine-cut surfaces or *etchplains.*

**Escarpment**: an asymmetric hill ridge usually produced on dipping sedimentary rocks, especially if capped by a resistant horizon, but may also result from a fault – to produce a fault scarp or a fault-line scarp.

**Etchplain**: an extensive low-gradient surface, produced by a long period of extensive chemical weathering. Characteristic of humid tropical or subtropical environments.

**Eustacy, as in eustatic sea-level change**: variations in global sea level resulting from the different proportions of the world's water stored as glacial ice. Eustatic sea levels are relatively high during global interglacials (as now) and low during global glacials.

**Evaporite**: a chemically precipitated sediment produced by evaporation to dryness of a saline water body e.g., of an arid-region playa lake. Most commonly composed (in increasing order of solubility) of calcium carbonate, gypsum or halite.

## F

**Fan delta**: the subaqueous portion of a river delta of a river delivering large amounts of (usually) coarse sediment. Its structure includes near-horizontal beds at the top (topset), dipping beds on the delta front (foreset) and near-horizontal beds at the base (bottomset beds).

**Fanhead trench**: the incised channel in the upper (proximal) part of an alluvial fan.

**Fault gouge**: squeezed and fragmented rock along a fault.

**Fetch**: the extent of open water in front of a coast that affects the magnitude of wind-generated waves.

**Floodplain**: a flat plain adjacent to an alluvial channel, composed of alluvial sediments, and whose surface defines the limits of the alluvial channel.

**Foreset**: Sloping beds deposited at the front of an accretionary wedge.

**Freeze-thaw weathering**: the mechanical weathering of rock by ice forming within cracks or along bedding planes in the rock. On freezing, water expands as it turns into ice, thus creating stresses that weaken the rock, eventually fracturing it and yielding angular clasts.

# G

**Garnet**: a family of crystalline silicate minerals, involving various amounts of at least some of the following metallic elements: calcium, aluminium, magnesium, manganese, iron, and chromium. Common in relatively high-grade metamorphic rocks, also occurring in volcanic rocks when the magma has passed through and incorporated continental crustal material.

**Garnet-mica schist**: a relatively high-grade metamorphic rock, containing garnet.

**Glacial periods**: periods (of duration in tens of thousands of years) during the Pleistocene when global temperatures were significantly lower than today, sea levels were lower, and much more of the world's water was in the form of glacial ice, particularly on the higher latitude continental areas, but also lower latitude higher mountain ranges. A signal is preserved in the oxygen isotope ratios in sediments, which has now become the basis for Pleistocene chronology (*see* Appendix 3). There were also other glacial periods earlier in the geological past.

**Gneiss**: a high-grade, coarsely crystalline metamorphic rock, often characterized by mineral banding.

**Growth fold**: a fold that undergoes deformation during sedimentation.

**Groundwater calcrete**: see Calcrete.

**Gully**: two forms: (i) hillslope gully: eroding channel cut into a hillslope, (ii) linear incised channel on a valley floor.

**Gully erosion**: downslope culmination of erosion by *overland flow*, water and sediment normally fed from rilled sideslopes. Extensive gully erosion normally termed *'badland'* erosion.

**Gypkarst**: karstic features developed on gypsum as opposed to limestone terrain.

**Gypsum**: calcium sulphate ($CaSO_4$. $nH_2O$). Present in seawater and saline lakes, deposited as an evaporate deposit.

# H

**Halite**: sodium chloride ($NaCl$), the main 'salt' in seawater – only precipitated from extreme concentrations, or from total desiccation.

**Hornblende**: one of several common families of complex silicate igneous rock-forming minerals. Characteristic of intermediate to moderately acidic rocks.

**Hydration/dehydration**: weathering process involving the absorption into/loss of water from the chemistry of a material.

# I J

**Igneous rocks**: rocks formed by crystallization from magma, either within the crust as intrusive rocks or at the surface as volcanic rocks.

**Ignimbrite**: a fine-grained pumice-like pyroclastic rock, formed by deposition from a turbulent pyroclastic flow.

**Imbrication**: the stacking of pebbles during deposition with their flatter surfaces dipping upstream.

**Incised meander:** a meandering river channel that is trenched into bedrock.

**Indurated**: hardened. When applied to geology, sedimentology, or geomorphology, relates to the transformation from sediment to rock by compaction or cementation, to create an erosionally resistant surface or layer.

**Infiltration capacity**: the rate at which the soil can absorb water by infiltration: varies with soil moisture, soil particle-size and pore space. Overland flow (surface runoff) can only occur if rainfall intensity exceeds infiltration capacity.

**Interglacial**: the periods between *glacials* – our present interglacial, the Holocene, has lasted for the last 10,000 years or so.

**Intersection point (on an alluvial fan)**: the point (usually in midfan) where an incised channel within a fan head trench, at a lower gradient than the fan surface, intersects the fan surface. From that point downfan, deposition is likely to be unconfined.

**Interstratal karst**: the erosion of (weak, perhaps marly) material from within a cave system within limestone or gypsum, as opposed to the removal of the limestone or gypsum by solution. Both processes may occur at the same time.

**Inverted relief**: relief where high topography coincides with structural lows and vice versa, e.g., synclinal ridges or anticlinal valleys.

# K L

**Karst**: a term coined from the 'Karst' region of former Yugoslavia to describe the geomorphology of solutional terrain, particularly of limestone terrain.

**Knickpoint**: a step or a break in the longitudinal profile of a stream caused by incision working its way upstream.

**Lacustrine**: related to a lake.

**Lahar**: a rapidly moving hot, wet volcanic mudflow or debris flow – highly dangerous. May be preserved in volcanic sedimentary sequences as a volcaniclastic deposit.

**Limestone**: a rock formed predominantly of calcium carbonate ($CaCO_3$).

**Lineaments**: linear patterns picked out at the land surface, by topography or vegetation, that reflects an underlying structure – particularly a fault.

**Liquid limits**: the moisture content at which a fine-grained (usually clayey) sediment or soil drains, and flows as a liquid.

**Lithology**: rock type and characteristics.

**Longshore drift**: movement of sediment along the beach, produced by oblique wave approach, often producing shore-parallel depositional features such as spits.

# M

**Mantle**: the intermediate layer within the earth between the crust and the core.

**Marble**: metamorphosed limestone, of crystalline calcium carbonate ($CaCO_3$).

**Marine regression/transgression**: the advance or retreat of a marine shoreline as a result of rising or falling sea level.

**Marl**: calcareous clay.

**Meandering channel**: single-thread alluvial channels, whose plan view comprises a series of bends alternating from side to side.

**Mechanical weathering**: the breakdown of rock by mechanical means.

**Metacarbonate**: a term used to describe metamorphosed limestone.

**Metagranite**: a term used to describe metamorphosed granite.

**Metamorphic rocks**: rocks (originally igneous or sedimentary rocks) whose properties (physical and/or chemical) have been radically altered by heat and/or pressure.

**Mica-schist**: a high-grade metamorphic rock, including abundant mica; may also contain other minerals, when it is prefaced as, for example: *garnet, graphite, hornblende, mica-schist.*

**Milankovitch cycles**: cyclic variations in the Earth's orbital characteristics, including a 96,000 year eccentricity cycle, a 40,000 year obliquity cycle and a 21,000 year precessional cycle, described by Milankovitch in the 1920s. Though Quaternary climatic fluctuations can be shown to bear some relationship with the combined effects of these cycles, the exact mechanism whereby these cycles might influence global climates is still uncertain.

**Misfit (underfit) stream**: (generally) a modern meandering alluvial river channel set in a much larger meandering valley, the implication being that as the geometry of the modern meanders is adjusted to modern flow conditions, the much larger valley meanders would be adjusted to much larger discharges in the past. There is a flaw to this argument in that the modern meanders are mobile meanders set in an alluvial floodplain and free to adjust by both erosion and deposition, whereas valley meanders are erosional forms, cut in bedrock and unable to adjust. They represent the cumulative effects of (admittedly) large flow events.

**Morphometry/Morphometrics**: measurements of the relief, areal extent, size, shape, slope angle or other geometric properties of landforms (e.g., slopes, drainage basins, river channels, alluvial fans, glacial features) and the numerical relationships between them.

**Mudslide**: a small-scale mass-movement phenomenon involving mud – an important process, characteristic of badlands and gully systems.

## N

**Nappe**: a forward-thrusted overfold produced during mountain building on a destructive plate boundary, characteristic of complex mountain chains such as the Alps.

**Neotectonic**: ongoing tectonic activity, continuing after the main mountain-building phase. In 'Alpine' young fold mountains this normally implies tectonics continuing into the Neogene.

**Normal fault**: a fault resulting from tension/extension, where the fault plane is usually fairly steep and dipping in the direction of downfault movement.

## O

**Oceanic crust**: the denser crust beneath the ocean basins, composed of basaltic rock. Formed at a constructive plate margin, usually a mid-oceanic ridge.

**Offlap**: deposition of successive sedimentary units away from their previous margins; may be induced by tectonic uplift of a land area or by falling sea levels.

**Olivine**: a ferromagnesium silicate mineral family, characteristic of basic igneous rocks.

**Order, as in stream order**: a system first devised by Horton, modified by Strahler, of classifying stream segments according to their position in the hierarchy of branching, so that unbranched headwater streams are 1st order streams, two 1st order streams join to form a 2nd order segment, two 2nd order segments to form a 3rd order segment, and so on.

**Overland flow**: surface runoff generated by rainfall excess or saturation overland flow.

**Oxidation/reduction**: chemical processes important in the weathering of rocks and soil formation by the addition/loss of oxygen. Particularly important in iron-rich environments, involving transformations between ferric and ferrous iron compounds.

# P

**Palaeochannel**: a former channel, now no longer functional.

**Palaeocurrent**: former current direction, preserved in sedimentary structures.

**Palaeosol** a fossil or buried soil.

**Palustrine**: related to an ephemeral lake, or to a swamp.

**Pediment**: a low-angle rock surface, created by scarp retreat, often concave in form. Pediments commonly occur at two scales: at a large (kilometre) scale related to the retreat of major slopes, and at a much smaller (micro-pediment) scale within actively eroding badlands.

**Pediplain**: extensive coalescent *pediments*: forming an erosion surface produced by the parallel retreat of hillslopes/escarpments.

**Pedogenic**: related to soil development. In the context of calcrete, relates to development through soil processes, rather than by 'groundwater' processes.

**Peneplain**: an extensive low-gradient surface, supposedly related to former stable base levels: the end product of the cycle of erosion. Nowadays this is rather a disputed concept.

**Periglacial**: a non-glaciated environment, but one subject to intense freeze-thaw action, commonly with permafrost.

**Phenocryst**: a larger crystal within a groundmass of smaller crystals.

**Phyllite**: a moderate fine-grained metamorphic rock (somewhere between a slate and a schist), sometimes referred to as a spotted slate).

**Phytokarst**: karstic dissolution accelerated by acid-secreting algae – common on tropical or sub-tropical limestone coasts or on exposed coral rock on such coasts.

**Pipe erosion/Piping**: subsurface erosion of tunnels, especially important in badland areas.

**Planar landslide**: a landslide where the surface of movement is planar downslope.

**Plastic limits**: the moisture content at which a fine-grained (usually clayey) sediment or soil deforms under its own weight by plastic flowage.

**Playa**: [not a Spanish beach!] a dry ephemeral lake in an arid environment; often surfaced by evaporate salts.

**Progradation**: deposition extending forwards in the flow direction.

**Provenance**: source area and pathways of sediment movement.

**Pyrite**: iron sulphide $(FeS_2)$.

**Pyroclastic rocks**: volcanic breccias: fragmented rocks, produced during volcanic eruptions.

## Q R

**Quartz**: silicon dioxide $(SiO_2)$ – a common rock-forming mineral, abundant in acid igneous rocks. Physically strong and almost chemically inert, therefore it persists through weathering cycles to be abundant in sedimentary rocks. The primary constituent of sand, and therefore of sandstone.

**Raised beach**: former beach deposits, related to a previous higher sea level and preserved above modern sea levels.

**Reef platform**: horizontal surface, created by outward coral reef growth near the contemporaneous sea level.

**Reef talus slope**: fragments of the coral reef platform that collapse onto the frontal reef slope, usually to be covered by further outward reef growth.

**Regolith**: superficial cover of weathered material and soil.

**Rejuvenation**: the response of a river system to uplift or a fall in base level, involving incision and the steepening of river profiles.

**Reverse fault**: A fault resulting from compression, usually with a steeply inclined fault plane (*see also* Thrust fault).

**Rhyolite**: lava of acid (granitic) composition.

**Rill erosion**: erosion on hillslopes by overland flow, where sheetflow begins to winnow out shallow channels that have more or less the same gradient as the slope itself. These channels may be ephemeral. These channels converge downslope into gullies, channels that are capable of eroding the underlying substrate and have a gradient markedly less than the slope gradient.

**River/stream capture**: aggressive drainage cuts back and intercepts an earlier line of drainage, diverting the drainage into the new course. The 'beheaded' stream loses its headwaters. The captor stream now becomes the major stream. Upstream the whole system may undergo *rejuvenation* and incision, due to a lower local *base level*.

**River terrace**: sediments of a former river, forming a flat surface (the former floodplain) perched above the modern river. A series of terraces may form a staircase of flat surfaces.

**Rockfall**: mass movement of rock fragments, generally from a cliff.

**Rotational failure**: landslide where the failure surface is arcuate.

# S

**Schist**: a high-grade, relatively fine-grained metamorphic rock.

**Sediment cascade**: the movement of sediment through the sediment system, usually from hillslopes through the fluvial system to the coastal system.

**Sedimentary rocks**: rocks formed by the accumulation and lithification of materials (sediments) that have passed through a cycle of weathering, erosion, transport and deposition. Includes clastic sedimentary rocks (e.g. sandstones) composed of fragments, and chemical precipitates.

**Seismite**: a fragmented breccia, produced by mass movement (rockfall?) triggered by earthquake activity.

**Sinuosity**: in a meandering river channel, the ratio between channel distance and the straight line distance between two points along the channel. Rather arbitrarily, channels with a sinuosity >1.5 are deemed to be meandering.

**Slickensides**: grooves along a fault plane indicative of the sense of movement along the fault.

**Slope failure**: landslide.

**Slump folds**: folds within a sedimentary rock; may be within an individual bed, induced by gravity rather than directly by tectonic forces;

rubbly breccia. May include phenocrysts of olivine or augite and other minerals related to hydrothermal alteration.

**Volcanic neck**: former volcanic vent, preserved as a lava plug, often forming an upstanding relief feature due to differential erosion of the (weaker) surrounding country rock (though the reverse may take place, where the former volcano forms a hollow).

**Volcaniclastic rocks**: rocks formed in volcanic areas, composed of fragments of volcanic material.

**Wave-cut platform (rock platform)**: a subhorizontal coastal erosional surface, sloping gently seawards.

**Weathering**: the mechanical and chemical breakdown of rock, yielding rock fragments (clasts), chemically altered rock, and ions that are removed in solution. A prerequisite for the operation of the *sediment cascade*.

# Appendix 5

# Logistics – how to get there, where to stay, how to get around

## How to get there

### Driving?

Remember it is a long way from the rest of Europe – even within Spain, Almeria is more than 600km from Madrid and even further away from Barcelona. So it is feasible to drive there from other parts of Spain, but requires a long time commitment from anywhere else in Europe.

### By rail

Almeria is a long way from the Spanish high-speed-train network. There are two daily SLOW trains from Madrid, and a few locals from Granada and other parts of Andalucia. The nearest reasonable point on a modern rail network is Alicante, accessible from both Madrid and Barcelona.

### By air

Almeria airport is served by Iberia flights from Madrid and Barcelona. It is also served by low cost airline Easyjet from a few airports in the UK (currently Gatwick and Manchester). Alicante has both regular flights and numerous flights by low-cost airlines from numerous UK airports, also from other selected airports elsewhere in Europe. Hiring a car there and driving south may be the best option; driving time to the field area is less than three hours. There are other airports (and car-hire facilities) at Malaga, Granada, and Murcia.

## Where to stay

In addition to the extensive tourist accommodation available at Mojacar and Vera Playa (near Garrucha) there are smaller-scale tourist facilities available near Carboneras, Agua Armarga and San Jose. There is hotel accommodation in Almeria and inland at Sorbas and near Tabernas. There are field centres at Urra, near Sorbas (currently closed, expected to reopen in 2015 or 2016), one in Lucainena , and one at Cortijo del Saltador in Rambla Honda (near Lucainena).

## How to get around

In addition to car-hire facilities at all the airports mentioned above, there are car-hire companies based in Mojacar. Shop around: prices are competitive!

# Index

Page numbers in *italic* denote illustrations. Page numbers in **bold** denote terms defined in glossary.

Index

# Index